ADVANCES IN NETWORK AND DISTRIBUTED SYSTEMS SECURITY

IFIP - The International Federation for Information Processing

IFIP was founded in 1960 under the auspices of UNESCO, following the First World Computer Congress held in Paris the previous year. An umbrella organization for societies working in information processing, IFIP's aim is two-fold: to support information processing within its member countries and to encourage technology transfer to developing nations. As its mission statement clearly states,

IFIP's mission is to be the leading, truly international, apolitical organization which encourages and assists in the development, exploitation and application of information technology for the benefit of all people.

IFIP is a non-profitmaking organization, run almost solely by 2500 volunteers. It operates through a number of technical committees, which organize events and publications. IFIP's events range from an international congress to local seminars, but the most important are:

- The IFIP World Computer Congress, held every second year;
- open conferences;
- working conferences.

The flagship event is the IFIP World Computer Congress, at which both invited and contributed papers are presented. Contributed papers are rigorously refereed and the rejection rate is high.

As with the Congress, participation in the open conferences is open to all and papers may be invited or submitted. Again, submitted papers are stringently refereed.

The working conferences are structured differently. They are usually run by a working group and attendance is small and by invitation only. Their purpose is to create an atmosphere conducive to innovation and development. Refereeing is less rigorous and papers are subjected to extensive group discussion.

Publications arising from IFIP events vary. The papers presented at the IFIP World Computer Congress and at open conferences are published as conference proceedings, while the results of the working conferences are often published as collections of selected and edited papers.

Any national society whose primary activity is in information may apply to become a full member of IFIP, although full membership is restricted to one society per country. Full members are entitled to vote at the annual General Assembly, National societies preferring a less committed involvement may apply for associate or corresponding membership. Associate members enjoy the same benefits as full members, but without voting rights. Corresponding members are not represented in IFIP bodies. Affiliated membership is open to non-national societies, and individual and honorary membership schemes are also offered.

ADVANCES IN NETWORK AND DISTRIBUTED SYSTEMS SECURITY

IFIP TC11 WG11.4
First Annual Working Conference on Network Security
November 26–27, 2001, Leuven, Belgium

Edited by

Bart De Decker
Katholieke Universiteit Leuven, DistriNet
Belgium

Frank Piessens
Katholieke Universiteit Leuven, DistriNet
Belgium

Jan Smits
Technische Universiteit Eindhoven
The Netherlands

Els Van Herreweghen
IBM Research Laboratory, Zürich
Switzerland

SPRINGER SCIENCE+BUSINESS MEDIA, LLC

Library of Congress Cataloging-in-Publication Data

A C.I.P. Catalogue record for this book is available from the Library of Congress.

DOI 10.1007/978-0-306-46958-9

Originally published by Kluwer Academic Publishers in 2002
MyCopy version of the original edition 2002

Printed on acid-free paper.
www.springer.com/mycopy

CONTENTS

Part Two - Invited Paper

PREFACE

The first Annual Working Conference of WG11.4 of the International Federation for Information Processing (IFIP), focuses on various state-of-the-art concepts in the field of Network and Distributed Systems Security.

Our society is rapidly evolving and irreversibly set on a course governed by electronic interactions. We have seen the birth of e-mail in the early seventies, and are now facing new challenging applications such as e-commerce, e-government, ... The more our society relies on electronic forms of communication, the more the security of these communication networks is essential for its well-functioning. As a consequence, research on methods and techniques to improve network security is of paramount importance.

This Working Conference brings together researchers and practioners of various disciplines, organisations and countries, to discuss the latest developments in security protocols, secure software engineering, mobile agent security, e-commerce security and security for distributed computing.

We are also pleased to have attracted two international speakers to present two case studies, one dealing with Belgium's intention to replace the identity card of its citizens by an electronic version, and the other discussing the implications of the security certification in a multinational corporation.

This Working Conference should also be considered as the kick-off activity of WG11.4, the aims of which can be summarized as follows:

- to promote research on technical measures for securing computer networks, including both hardware- and software-based techniques.
- to promote dissemination of research results in the field of network security in real-life networks in industry, academia and administrative institutions.

- to promote education in the application of security techniques, and to promote general awareness about security problems in the broad field of information technology.

Researchers and practioners who want to get involved in this Working Group, are kindly requested to contact the chairman. More information on the workings of WG11.4 is available from the official IFIP-website: `http://www.ifip.at.org/`.

Finally, we wish to express our gratitude to all those who have contributed to this conference in one way or another. We are grateful to the international referee board who reviewed all the papers and to the authors and invited speakers, whose contributions were essential to the success of the conference. We would also like to thank the participants whose presence and interest, together with the changing imperatives of society, will prove a driving force for future conferences to come.

PROF. B. DE DECKER

ACKNOWLEDGEMENTS

Organised by:

K.U.Leuven, Dept. of Computer Science, DistriNet
IFIP/TC-11 Working Group 11.4 (Network Security)

Supported by:

Scientific Research Network on "Foundations of Software Evolution", and as such, partially financed by the Fund for Scientific Research - Flanders (Belgium)

Financially Supported by:

IBM Research
Telindus
Ubizen
Utimaco Safeware Belgium

Programme Committee:

Bart De Decker, (chair), K.U.Leuven, Belgium
Jan M. Smits, (co-chair), T.U.Eindhoven, The Netherlands
Els Van Herreweghen, (co-chair), IBM Research Lab, Zurich, Switzerland

William J Caelli, Queensland Univ. of Technology, Australia
Herve Debar, France Telecom R&D, France
Serge Demeyer, Univ. of Antwerp, Belgium
Yves Deswarte, LAAS-CNRS, Toulouse, France
Jan Eloff, Rand Afrikaans Univ., South Africa
Dimitris Gritzalis, Athens Univ. of Economics & Business, Greece
Manfred Hauswirth, Technical Univ. of Vienna, Austria
Andrew Hutchison, MGX Consulting, South Africa
Guenter Karjoth, IBM Zurich Research Lab, Switzerland
Kwok-Yan Lam, PrivyLink International Limited, Hong Kong
Marc Lobelle, UCL, Belgium
Keith Martin, Royal Holloway, Univ. of London, United Kingdom
Refik Molva, Institut Eurécom, France
Frank Piessens, K.U.Leuven, Belgium

Hartmut Pohl, Univ. of Applied Sciences Bonn-Rhein-Sieg, Köln, Germany
Reinhard Posch, Graz Univ. of Technology, Austria
Bart Preneel, K.U.Leuven, Belgium
Kai Rannenberg, Microsoft Research Cambridge, United Kingdom
Peter Ryan, Norwegian Computing Center, Oslo, Norway
Pierangela Samarati, Univ. of Milan, Italy
Einar Snekkenes, Norsk Regnesentral, Oslo, Norway
Henk van Tilborg, T.U.Eindhoven, The Netherlands
Vijay Varadharajan, Macquarie Univ., Australia
Basie Von Solms, Rand Afrikaans Univ., South Africa
Rossouw Von Solms, Port Elizabeth Technikon, South Africa
Jozef Vyskoc, VaF, Bratislava, Slovak Republic
Tatjana Welzer, Univ. of Maribor. Slovenia

Reviewers:

Bussard, Laurent, France
Debar, Hervé, France
De Decker, Bart, Belgium
De Win, Bart, Belgium
Demeyer, Serge, Belgium
Deswarte, Yves, France
Družovec, Marjan, Slovenia
Eloff, Jan, South Africa
Gritzalis, Dimitris, Greece
Hauswirth, Manfred, Austria
Hutchison, Andrew, South Africa
Karjoth, Guenter, Switzerland
Lam, Kwok-Yan, Hong Kong
Lobelle, Marc, Belgium
Martin, Keith, United Kingdom
Molva, Refik, France
Piessens, Frank, Belgium
Pohl, Hartmut, Germany
Posch, Reinhard, Austria
Preneel, Bart, Belgium
Rannenberg, Kai, United Kingdom
Ryan, Peter, Norway
Samarati, Pierangela, Italy

Schoenmakers, Berry, The Netherlands
Smits, Jan, The Netherlands
Snekkenes, Einar, Norway
Van Herreweghen, Els, Switzerland
Vanhaute, Bart, Belgium
van Tilborg, Henk, The Netherlands
Varadharajan, Vijay, Australia
Von Solms, Basie, South Africa
Von Solms, Rossouw, South Africa
Vyskoc, Josef, Slovak Republic
Welzer, Tatjana, Slovenia

PART ONE

Reviewed Papers

A ROLE-BASED SPECIFICATION OF THE SET PAYMENT TRANSACTION PROTOCOL

Hideki Sakurada
NTT Communication Science Laboratories,
NTT Corporation,
3-1 Morinosato-Wakamiya, Atsugi, Kanagawa, 243-0198 Japan
sakurada@theory.brl.ntt.co.jp

Yasuyuki Tsukada
NTT Communication Science Laboratories,
NTT Corporation,
3-1 Morinosato-Wakamiya, Atsugi, Kanagawa, 243-0198 Japan
tsukada@theory.brl.ntt.co.jp

Abstract In this paper, we define a language for specifying security protocols concisely and unambiguously. We use this language to formally specify the protocol for payment transactions in Secure Electronic Transaction (SET), which has been developed by Visa and MasterCard.

In our language, a protocol is specified as a collection of processes. Each process expresses the role of a participant. In the role-based specification, the components that a participant sees in a message can be stated explicitly. This is important in specifying protocols like that for the SET payment transactions because in such protocols some message components are encrypted and invisible to some participants.

We simplify the SET payment transaction protocol into the exchanges of six messages. Because our future goal is to formally analyze the security properties that Meadows and Syverson discussed, we make the simplified protocol contain the parameters used in their security properties. And we also refrain from excessive simplification. For example, we use dual signature in the payment request message as it is specified in the SET specification books, while most of the other works do not use it. Our specification can serve as a starting point for a formal analysis of the protocol.

Keywords: Formal methods, security protocols, electronic commerce

1. Introduction

Security protocols are used in distributed systems to protect the secrecy of messages and to identify users. It is well known that designing them is an error-prone task. The most significant issues concerning security protocols are that (1) attacks on them may succeed even without breaking the cryptographic algorithms used and that (2) it may be difficult to make sure of the correctness of a small protocol that involves exchanges of only a few messages. Some examples of protocol failures are presented in (Anderson and Needham, 1995; Clark and Jacob, 1997).

Formal methods can be used to analyze security protocols. With the methods, protocols are specified and their security properties are verified. Indeed, many formal methods have been developed (Meadows, 1996; Paulson, 1998; Denker et al., 2000) and succeeded in finding errors in protocols or verifying their correctness (Burrows et al., 1990; Paulson, 1998). However, it is hard to apply these methods to large protocols. This is because large protocols are complex and there are no appropriate tools for analyzing such complex protocols. With a tool designed for small protocols, specifying complex protocols and their security properties is hard. Moreover, the obtained specifications tend to be lengthy and unintuitive. To avoid these difficulties, protocols are usually simplified and the simplified protocols are verified instead.

In this paper, we discuss the Secure Electronic Transaction (SET) protocol (SET Secure Electronic Transaction LLC, 1997a; SET Secure Electronic Transaction LLC, 1997b; SET Secure Electronic Transaction LLC, 1997c). In particular, we formally specify the payment transaction protocol that is a part of SET. This formal specification serves as a starting point of a formal analysis of the protocol.

SET has been developed by Visa and MasterCard for secure electronic commerce using payment cards. Over six hundred pages are needed to explain and specify it. There are some works on the formal specification and the analysis of the protocol (Lu and Smolka, 1999; Bolignano, 1997; Kessler and Neumann, 1998). However, they simplified the protocol excessively in order to reduce the complexity. For example, most of these simplified protocols did not use dual signature, which is one of the characteristics of SET. Since we aim at verifying security properties that Meadows and Syverson discussed in (Meadows and Syverson, 1998), we include in our simplified protocol the parameters that occur in the properties. We also make the simplified protocol use dual signature. In order to describe the specification concisely and unambiguously, we first define a protocol specification language. In our language, a protocol is specified as a collection of processes that express the roles of the

$$
\begin{array}{lcl}
A & \rightarrow S : & A, B, N_A \\
S & \rightarrow A : & \{N_A, B, K_{AB}, \{K_{AB}, A\}_{K_{BS}}\}_{K_{AS}} \\
A & \rightarrow B : & \{K_{AB}, A\}_{K_{BS}} \\
B & \rightarrow A : & \{N_B\}_{K_{AB}} \\
A & \rightarrow B : & \{N_B - 1\}_{K_{AB}}
\end{array}
$$

Figure 1. A typical message flow in the Needham-Schroeder shared-key protocol

participants in the protocol. This is useful for describing the specification of the SET payment transaction protocol.

The rest of this paper is organized as follows. We first define a language for specifying large security protocols concisely and unambiguously (Section 2). We then use it to specify the SET payment transaction protocol (Section 3). We finally summarize our results and mention some related works (Section 4).

2. Protocol Specification Language

Before presenting our protocol specification language, we briefly explain our design policy for it.

Security protocols are often explained by showing a typical message flow. For example, a typical message flow of the Needham-Schroeder shared-key protocol (Needham and Schroeder, 1978) is shown in Figure 1. The first line means that a participant A sends a message composed of her name, the name of the participant she wants to authenticate, and a fresh nonce (random number) to the authentication server S. The second line means that S replies with message $\{N_A, B, K_{AB}, \{K_{AB}, A\}_{K_{BS}}\}_{K_{AS}}$ to A. This message is obtained by encrypting N_A, B, a newly generated key K_{AB} to be shared by A and B, and $\{K_{AB}, A\}_{K_{BS}}$ with the key K_{AS}. The $\{K_{AB}, A\}_{K_{BS}}$ is obtained by encrypting K_{AB} and A with the key K_{BS}. A, B, and N_A on the second line refer to themselves on the first line, respectively. Since A is assumed to know K_{AS} and is not assumed to know K_{BS}, she can decrypt $\{N_A, B, K_{AB}, \{K_{AB}, A\}_{K_{BS}}\}_{K_{AS}}$ and can not decrypt $\{K_{AB}, A\}_{K_{BS}}$.

Explanations by showing a typical message flow are concise and intuitive. However, they can not explicitly handle what each participant can see in a message because each line expresses the sending and receiving of a message at the same time. For example, on the second and the third line in the previous example, A receives a messages that includes

$$
\begin{array}{cc}
\{A, B, N_A\} & \rightarrow \\
\Downarrow & \\
\{N_A, B, K_{AB}, X\}_{K_{AS}} & \leftarrow \\
\Downarrow & \\
X & \rightarrow \\
\Downarrow & \\
\{N_B\}_{K_{AB}} & \leftarrow \\
\Downarrow & \\
\{N_B - 1\}_{K_{AB}} & \rightarrow
\end{array}
$$

Figure 2. The initiator role in the Needham-Schroeder shared-key protocol

$\{K_{AB}, A\}_{K_{BS}}$ and sends it to B. Without assumptions on the knowledge of A, it is not clear whether if she knows the content $\{K_{AB}, A\}$ of the message or not. This ambiguity may cause human-errors in specifying complex protocols that use cryptography frequently.

To avoid this problem, we specify a protocol as a collection of processes that express the roles of the participants in the protocol. To illustrate this, we show, in Figure 2, a process that are related to A's role in the previous example. Note that we use a variable X for the encrypted component in the message from S to A. It is clear that A sends the component X to B as it is.

Now we define our protocol specification language. Since we assume the Dolev-Yao (Dolev and Yao, 1981) model, we define the set of *messages* as an algebra made from participants' names, natural numbers (including nonces), and keys with tupling and cryptographic operations. The formal syntax of messages is as follows.

$$
\begin{array}{lll}
M ::= & A & \text{; participant's name} \\
\mid & K & \text{; key} \\
\mid & N & \text{; natural number} \\
\mid & \{M_1, \cdots, M_n\} & \text{; tuple} \\
\mid & \{M\}_K & \text{; encryption of message } M \text{ using key } K \\
\mid & \mathsf{H}(M) & \text{; hash of message } M
\end{array}
$$

H is a collision-free one-way hash function. We write K^{-1} for the decryption key of a key K. For example, $\{A, N_A\}_K$ is a message obtained by encrypting a tuple of A and N_A with K, where A, N_A, and K are an participant's name, a nonce, and a key, respectively.

Since our language has variables, we define the set of *terms* by extending the previous syntax with variables.

$$
\begin{array}{lll}
T & ::= & \cdots \\
& | & X \quad \text{; variable whose name is } X
\end{array}
$$

Because we usually use variables instead of concrete names, nonces, and keys, we regard A, N_A, K, etc. that occur in terms as variables unless otherwise noted explicitly.

We finally define the set of *processes* with the following syntax. We specify a protocol as the set of processes of its participants.

$$
\begin{array}{llll}
P & ::= & \mathsf{End} & \text{; silent process} \\
& | & \mathsf{Send}\ T\ P & \text{; sending of message } T \\
& | & \mathsf{Recv}\ T\ P & \text{; receiving of message } T \\
& | & \mathsf{New}\ X\ P & \text{; generating of a fresh nonce } X \\
& | & \mathsf{Let}\ X = T\ P & \text{; binding of } T \text{ to the local variable } X \\
& | & \mathsf{Assert}\ Q\ P & \text{; checking of proposition } Q
\end{array}
$$

We don't specify the receiver and the sender of a message in Send TP and Recv TP, respectively because we assume that there exist intruders that can capture any message on networks and can send any message they can construct. We understand that a process of the form New X P binds free occurrences of X in P. In other words, in a process New X P, the variables X that occur in P refer to the newly generated nonce X. We also understand that a process of the form Recv T P does pattern-matching and variable-binding. For example, a process Recv $\{N, \mathsf{H}(N)\}$ P accepts $\{2001, \mathsf{H}(2001)\}$, where variable N is bound to the number 2001. The process however does not accept $\{2001, \mathsf{H}(2002)\}$.

Assert Q P acts as P if proposition Q holds, otherwise it acts as End. The set of propositions depends on the system used for analysis. Since we use Isabelle (Paulson, 1994), a proof checker of higher-order logics, we can use any proposition in Isabelle.

As an example, we specify the role of A, the initiator, in the Needham-Schroeder shared-key protocol in Figure 3. The process is parametrized by her name A, the responder's name B, and the key K_{AS}.

3. A Specification of the SET Payment Transaction Protocol

In this section, we give a formal specification of the SET payment transaction protocol. Since our future goal is to verify security properties that include those which Meadows and Syverson discussed in (Meadows and Syverson, 1998), we simplify the protocol into the exchanges of six

$initiator(A, B, K_{AS}) =$
New N_A
Send $\{A, B, N_A\}$
Recv $\{N_A, B, K_{AB}, X\}_{K_{AS}}$
Send X
Recv $\{N_B\}_{K_{AB}}$
Send $\{N_B - 1\}_{K_{AB}}$
End

Figure 3. The process that specifies the initiator role of the Needham-Schroeder shared-key protocol

messages that include the parameters used in their security properties. Meadows and Syverson developed a method to describe security properties flexibly and discussed the security properties that the payment transaction protocol is expected to satisfy. However, they did not specify the protocol formally. Our formal specification is needed in order to verify the security properties. We also make the simplified protocol use dual signature, which is one of the characteristics of the original protocol.

We first overview the SET payment transaction protocol. Three parties, a cardholder, a merchant, and a payment gateway, are involved in a payment transaction in SET. This protocol is invoked after the cardholder has completed browsing, selection, and ordering. One of the purposes of the protocol is to securely send the payment information, which includes the account number of the payment-card of the cardholder and the amount of money that he will pay for the order, to the payment gateway.

A typical message flow of the protocol is shown in Figure 4. We show only the six messages that our simplified protocol has. We also omit the structures of the messages in the figure. The cardholder and the merchant first exchange the identifiers of the transaction in *PInitReq* and *PInitRes* messages. The identifiers are referred to in subsequent messages. The cardholder then sends the purchase request message *PReq* to the merchant. This message includes the amount of money that the cardholder will pay and her payment-card number. She keeps the number secret from the merchant by encrypting a component that includes it. The merchant sends the gateway *AuthReq* message that includes the component. The gateway checks the validity of the payment-card number, processes the payment, and returns the result to the merchant in

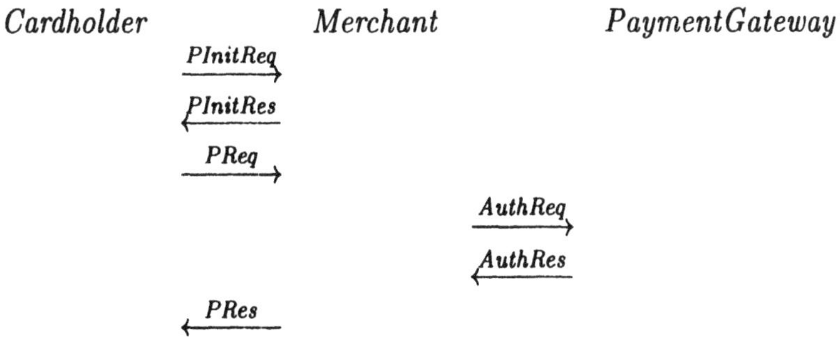

Figure 4. A typical message flow in the SET payment transaction protocol

$$
\begin{aligned}
\mathsf{L}(M_1, M_2) &= \{M_1, \mathsf{H}(M_2)\} \\
\mathsf{SO}(A_s, M) &= \{\mathsf{H}(M)\}_{\mathsf{SK}(A_s)} \\
\mathsf{S}(A_s, M) &= \{M, \mathsf{SO}(A_s, M)\} \\
\mathsf{E}(A_r, M, K) &= \{\{M\}_K, \{K^{-1}\}_{\mathsf{EK}(A_r)}\} \\
\mathsf{EX}(A_r, M_1, M_2, K) &= \{\{\mathsf{L}(M_1, M_2)\}_K, \{K^{-1}, M_2\}_{\mathsf{EK}(A_r)}\} \\
\mathsf{Enc}(A_s, B_r, M, K) &= \mathsf{E}(B_r, \mathsf{S}(A_s, M), K) \\
\mathsf{EncB}(A_s, B_r, M_1, M_2, K) &= \{\mathsf{Enc}(A_s, B_r, \mathsf{L}(M_1, M_2), K), M_2\}
\end{aligned}
$$

Figure 5. Operations on messages used in the SET payment transaction protocol

AuthRes message. The merchant receives it and sends the result to the cardholder in *PRes* message.

Various cryptographic operations are used in SET. We define each of the operations used in our protocol as a function on the set of messages in our language. The definitions are essentially the same as what Bella et al. did in their verification of the SET cardholder registration protocol (Bella et al., 2000). We show the definitions in Figure 5. The subscripts r and s of names of participants indicate that the participants appear as the receiver and the sender of a message, respectively. $\mathsf{L}(M_1, M_2)$ contains a linkage from message M_1 to message M_2. $\mathsf{SO}(A_s, M)$ is the signature of a participant A_s on message M. $\mathsf{S}(A_s, M)$ is message M with the signature of A_s. Enc models a signed-then-encrypted message. EncB models a signed-then-encrypted message with an external baggage.

$Cardholder(C, M, P, OD, PurchAmt, PAN, PANSecret) =$
New $RRPID_1$
New LID_C
New $Chall_C$
// *PInitReq*
Send $\{RRPID_1, LID_C, Chall_C\}$
// *PInitRes*
Recv $\mathsf{S}(M, \{\{LID_C, LID_M, XID\}, RRPID_1, Chall_C, Chall_M\})$
Let $TransID = \{LID_C, LID_M, XID\}$
New $ODSalt$
New $RRPID_2$
Let $PANData = \{PAN, PANSecret\}$
Let $PIHead = \{TransID, \mathsf{H}(OD), PurchAmt\}$
Let $OIData = \{TransID, RRPID_2, Chall_C, \mathsf{H}(OD), ODSalt\}$
Let $PIData = \{PIHead, PANData\}$
New K
// *PReq*
Send $\{ \{ \mathsf{SO}(C, \{\mathsf{H}(PIData), \mathsf{H}(OIData)\}),$
$\mathsf{EX}(P, \mathsf{L}(PIHead, OIData), PANData, K)\},$
$\{OIData, \mathsf{H}(PIData)\}\}$
// *PRes*
Recv $(\mathsf{S}(M, \{TransID, RRPID_2, Chall_C\}))$

Figure 6. The cardholder process in the SET payment transaction protocol

EK and SK are the functions that relate each participant to his public encryption key and his public signature key, respectively.

The processes of a cardholder, a merchant, and a payment gateway are shown in Figures 6, 7 and 8, respectively.

Here, C, M, and P are the names of a cardholder, a merchant, and a payment gateway, respectively. OD, PAN, $PurchAmt$ and $AuthReqAmt$ are an order description, the account number of a payment-card, the amount of money that a cardholder will pay, and the amount of money that a merchant requires, respectively. $PANSecret$ is used to prevent guessing attacks on PAN. $ValidPANSet$ is the set of valid PANs. It does not appear in the SET specification books. We introduce it to model the authentication of payment-cards. Dual signature is used in the *PReq* message. The message is composed of the following three parts: $\mathsf{SO}(C, \{\mathsf{H}(PIData), \mathsf{H}(OIData)\})$, $\mathsf{EX}(P, \mathsf{L}(PIHead, OIData), PANData, K)$ and $\{OIData, \mathsf{H}(PIData)\}$.

```
Merchant(M, C, P, OD, ODSalt, AuthReqAmt) =
  // PInitReq
  Recv {RRPID1, LIDC, ChallC}
  New LIDM
  New XID
  New ChallM
  Let TransID = {LIDC, LIDM, XID}
  // PInitRes
  Send S(M, {TransID, RRPID1, ChallC, ChallM})
  Let OIData = {TransID, RRPID2, ChallC, H(OD), ODSalt}
  // PReq
  Recv {{SO(C, HPIData, H(OIData)), PIBody},
        {OIData, HPIData}}
  New RRPID3
  New K1
  // AuthReq
  Send EncB( M, P, {RRPID3, TransID, AuthReqAmt},
             {SO(C, {HPIData, H(OIData)}), PIBody}, K1)
  // AuthRes
  Recv Enc(P, M, {RRPID3, TransID, AuthAmt}, K2)
  Assert AuthReqAmt = AuthAmt
  // PRes
  Send S(M, {TransID, RRPID2, ChallC})
```

Figure 7. The merchant process in SET payment transactions

```
Gateway(P, C, M, ValidPANSet) =
  // AuthReq
  Recv EncB( M, P, {RRPID3, TransID, AuthReqAmt}
             { SO(C, { H({{TransID, HOD, PurchAmt}, PANData}),
                       HOIData}),
               EX(P, { {TransID, HOD, PurchAmt},
                       HOIData}, PANData, K1)})
  Assert PANData ∈ ValidPANSet
  Assert PurchAmt = AuthReqAmt
  Let AuthAmt = AuthReqAmt
  New K2
  // AuthRes
  Send Enc(P, M, {RRPID3, TransID, AuthAmt}, K2)
```

Figure 8. The gateway process in SET payment transactions

The second part cannot be decrypted by a merchant and should be passed to a payment gateway. The content of the third part should be read by a merchant. The first part is the signature on $\{\mathsf{H}(PIData), \mathsf{H}(OIData)\}$. A participant who receives either of the last two parts can compute $\{\mathsf{H}(PIData), \mathsf{H}(OIData)\}$ and can check the signature.

4. Concluding Remarks

In this paper, we have defined a language for specifying security protocols and have used it to formally specify the SET payment transaction protocol. In our language, a security protocol is specified as a collection of processes. Each process defines the role of a participant. This is useful in specifying complex protocols concisely and unambiguity.

We have simplified the SET payment transaction protocol and have specified it formally. We aim at verifying various security properties of the protocol including those that Meadows and Syversion discussed in (Meadows and Syverson, 1998). Our specification can serve as a starting point for a formal analysis that take into account dual signature of the protocol.

We have already implemented our specification language on the Isabelle theorem prover (Paulson, 1994) and have written the specification in it. We are also developing a protocol execution model and a language to describe security properties concisely. In the execution model, a state of a participant is modeled as a process in our language and an environment, a set of variable-value pairs. The environment corresponds to the data that the participant uses. For example, in a key exchange protocol, the environment of a participant may include the name of the agent that a participant will talk with and the key she will exchange. The environments can also be used to describe security properties concisely. In the previous example, the agreement between the participants about the key can be expressed as coincidence between parts of the environments of participants. We plan to describe security properties that the SET payment transaction protocol should satisfy in our language and to verify them. We further have to make clear the correspondence between the original payment transaction protocol used in actual e-commerce and the simplified version we presented in this paper.

We finally mention some related works. There are a lot of works applying formal methods to protocol analyses. We will mention a few languages used to specify protocols in these works. CSP (Hoare, 1985) is used to specify security protocols in many protocol verification systems (Schneider, 1997; Roscoe, 1995). It seems that protocol specifications in our language can be easily translated into a collection of CSP processes and that tools for CSP can be used to verify the security.

Cervesato (Cervesato, 2001a; Cervesato, 2001b) proposed a protocol specification language, called Typed MSR. It is a kind of multiset rewriting system. His language also uses role-based descriptions. Protocol specifications in our language are more concise than those in his language because, in his language, predicates that correspond to the state of each participant must be explicitly written.

There are some works on security analyses of the SET protocol. Lu and Smolka (Lu and Smolka, 1999) formally specified the protocol as CSP processes and verified five correctness properties of the protocol using the FDR (Formal Systems Ltd, 1998) model checker. They however did not analyze dual signature and did not assume the existence of intruders in their analysis.

Meadows and Syverson (Meadows and Syverson, 1998) developed a security specification language for their protocol analyzer (Meadows, 1996). They also discussed the security properties that the SET payment transaction protocol is expected to satisfy. However, they did not give the specification of the protocol formally, and they left the actual verification of the security for future work. As far as we know, no result on the verification has been published yet. Our specification can serve as a starting point of a formal verification of security properties they discussed.

Bolignano (Bolignano, 1997) proposed a method to analyze security protocols. He took a protocol that resembles SET as an example. He has not completed the analysis of SET itself as far as we know.

Bella et al. (Bella et al., 2000) analyzed the cardholder registration protocol in SET. The protocol is used to exchange certificates needed in the payment transactions. They use the inductive method (Paulson, 1998) for their analysis.

Kessler and Neumann (Kessler and Neumann, 1998) defined a logic to treat the accountability of participants in electronic commerce protocols. They used their logic to analyze the accountability of a merchant in SET. They took into account dual signature, although they treated only the *PReq* message.

Acknowledgments

The authors thank Kazuo Ohta, Akira Takura, and Kiyoshi Shirayanagi for their helpful comments and encouragement.

References

Anderson, R. and Needham, R. (1995). Programming satan's computer. In *Computer Science Today: Recent Trends and Developments*, volume 1000 of *LNCS*, pages 426–440. Springer-Verlag.

Bella, G., Massacci, F., Paulson, L. C., and Tramontano, P. (2000). Formal verification of cardholder registration in SET. In *6th European Symposium on Research in Computer Security (ESORICS'00)*, volume 1895 of *LNCS*, pages 159–174. Springer-Verlag.

Bolignano, D. (1997). Towards the formal verification of electronic commerce protocols. In *10th IEEE Computer Security Foundations Workshop*, pages 133–146.

Burrows, M., Abadi, M., and Needham, R. (1990). A logic of authentication. *ACM Transactions on Computer Systems*, 8(1):18–36.

Cervesato, I. (2001a). Typed MSR: Syntax and examples. In *Information Assurance in Computer Networks: Methods, Models, and Architectures for Network Security (MMM-ACNS'01)*, volume 2052 of *LNCS*, pages 159–177. Springer-Verlag.

Cervesato, I. (2001b). Typed multiset rewriting specifications of security protocols. In *1s Irish Conference on the Mathematical Foundations of Computer Science and Information Technology (MFCSIT'00)*, ENTCS. Elsevier. To appear.

Clark, J. and Jacob, J. (1997). A survey of authentication protocol literature: Version 1.0. Technical report, Department of Computer Science, University of York.

Denker, G., Millen, J., and Rueß, H. (2000). The CAPSL integrated protocol environment. SRI Technical Report SRI-CSL-2000-02, SRI International.

Dolev, D. and Yao, A. C. (1981). On the security of public key protocols (extended abstract). In *22nd Annual Symposium on Foundations of Computer Science*, pages 350–357. IEEE.

Formal Systems Ltd (1998). FDR2 user manual.

Hoare, C. A. R. (1985). *Communicating Sequential Processes.* Prentice Hall.

Kessler, V. and Neumann, H. (1998). A sound logic for analysing electronic commerce protocols. In *5th European Symposium on Research in Computer Security (ESORICS'98)*, volume 1485 of *LNCS*, pages 345–360. Springer-Verlag.

Lu, S. and Smolka, S. (1999). Model checking SET Secure Electronic Transaction Protocol. In *7th International Symposium on Modeling, Analysis and Simulation of Computer and Telecommunication Systems (MASCOTS'99)*, pages 358–365. IEEE.

Meadows, C. (1996). The NRL protocol analyzer: an overview. *Journal of Logic Programming*, 26(2):113–131.

Meadows, C. and Syverson, P. (1998). A formal specification of requirements for payment transactions in the SET protocol. In *Financial Cryptography '98*, volume 1465 of *LNCS*, pages 122–140. Springer Verlag.

Needham, R. and Schroeder, M. (1978). Using encryption for authentication in large networks of computers. *Communications of the ACM*, 21(12):993–999.

Paulson, L. C. (1994). *Isabelle: A Generic Theorem Prover*, volume 828 of *LNCS*. Springer-Verlag.

Paulson, L. C. (1998). The inductive approach to verifying cryptographic protocols. *Journal of Computer Security*, 6(1):85–128.

Roscoe, A. W. (1995). Modelling and verifying key-exchange protocols using CSP and FDR. In *8th IEEE Computer Security Foundations Workshop*, pages 98–107.

Schneider, S. (1997). Verifying authentication protocols with CSP. In *10th IEEE Computer Security Foundations Workshop*, pages 3–17.

SET Secure Electronic Transaction LLC (1997a). SET secure electronic transaction book 1: Business description.

SET Secure Electronic Transaction LLC (1997b). SET secure electronic transaction book 2: Programmer's guide.

SET Secure Electronic Transaction LLC (1997c). SET secure electronic transaction book 3: Formal protocol definition.

INFORMATION SECURITY: MUTUAL AUTHENTICATION IN E-COMMERCE

S.H. Von Solms
Department of Computer Science
Rand Afrikaans University
PO Box 524, AUCKLAND PARK, 2006
South Africa
Tel: +27 11 489-2847   Fax: +27 11 489-2138
basie@rkw.rau.ac.za

M.V.Kisimov
Department of Computer Science
Rand Afrikaans University
Johannesburg, South Africa
Tel + 27 11 673-0163   Fax +27 11 673-0163
kisimov@yahoo.com

Abstract: Information Security is ever increasingly becoming an important topic when it comes to network communications. This greatly concerns areas of electronic commerce, especially online shopping and money transfers. This paper outlines a methodology for securing electronic communication between e-Merchants and online shoppers. The methodology is based on a simple hierarchy of a trusted third party and communicating hosts. The paper further explains how the new methodology avoids e-commerce pitfalls of current technologies and presents an approach for securing currently unsecured online shoppers, in the process of making them capable of performing safe and secure network transactions.

Keywords: Certification Authority, Authentication, Guideline, Security, Digital Certificates, Encryption, Asymmetric Cryptography, Digital Signature.

1. INTRODUCTION

In an ever-improving technological world, e-commerce is becoming an increasingly popular a tool for communication, business and analysis. Consequently the value of information being transmitted and its preservation is of high importance to its owner. This paper presents a new methodology, based on current technologies, which avoids security pitfalls to which current e-commerce standards are prone. This methodology deals with outlining a clear process for securing an average online shopper, with the necessary attributes needed for performing a safe online transaction. Further it defines an authentication process for verification of communicating parties' identities, using a trusted third party in the form of a Certification Authority (CA). As a result this methodology provides a legal process for creating nonrepudiation of performed transactions, which can be used in verifying the origin and the occurrence of a transaction.

1.1 Outline

Section two of this document looks at background work done to improve authentication between communicating parties as well as focusing on current electronic commerce problems and security loopholes. Section three of the document outlines the proposed methodology, which is the main focus of this document. Finally section four serves as a logical end to the document summarizing the important points made throughout.

2. BACKGROUND AND SECURITY PROTOCOLS

This section presents certain pitfalls of current e-commerce strategies and standards, in terms of security, customer satisfaction, authentication and technological standards. It will further present certain security weaknesses of the SSL protocol, which can be exploited by malicious parties. The points discussed here, present an obstacle to companies and individuals in establishing proper standards for electronic commerce and information security.

2.1 Customer Satisfaction

In a recently conducted study [PWC 98], statistics vital and worrying to corporations conducting business over the Internet as well as to online shoppers have emerged. The study showed that 60 percent of initiated online transactions are abandoned due to lack of online support, necessary security measures and lack of a standardised legal process for completing the online transactions. The ratio of completed to initiated transactions should be very discouraging to online merchants. Problems arising due to complex techniques and unproven technologies, often lead potential customers dropping transactions midway through and searching for different online merchants. E-Merchants, who present customers with long and extended processes for completing transactions, are usually the ones to suffer from lost business [PWC 98]. Security is of high concern to as many as 58 percent of online shoppers and only fewer than 10 percent of online shoppers are not concerned with security while performing a sensitive Internet transaction.

2.2 Online Digital Certificate Verification

Research performed by the authors reveals that many commercial products used for online transactions, which employs asymmetric cryptography and Digital Certificates (DCs) as method of encryption and authentication, over unprotected networks, do not provide methodology for online verification of these DCs. The need for such verification is based on the fact that Digital Certificates can be tampered with, corresponding private keys can be lost or compromised. This can cause information secured with these keys, to be compromised and to become volatile to malicious security attacks. Currently existing Certification Authorities (CAs) and PKIs such as VeriSign and Entrust [CTNS 00], [VS 01] implement special Certificate Revocation Lists (CRLs) [BPKIC 01], which hold a list of certificates, which are registered or issued by the CA or PKI. These lists represent DCs, which have been compromised in any manner. A verification of the DCs in use between communicating parties, in the issuing CA's CRL will confirm that in fact, these certificates have not been reported to be compromised. This can serve as a verification of the security of the data being transmitted. Such verification is not a property of any of the commercial products, which concern themselves with digital, network-based communication. Taking the problem further, if a certain certificate has been compromised, but the tampering has gone undetected to anyone, this certificate would not be reported to the CA and consequently not listed in the CA's CRL. This would leave any communication employing this DC compromised.

2.3 Authentication of online customers

Credit card fraud is a common occurrence for e-Merchants [PWC 98]. Reasons for fraud vary from lost credit cards, falsely generated credit information and duplicated or stolen credit cards being passed to the Merchant. Currently true authentication of the online shoppers is not always possible. Very few commercial or other products are in place, which deal with authentication of communicating parties over an open network. The latest version of the SSL protocol [SSL 96] provides for the possibility of such authentication. This however is not a prerequisite for the functionality of SSL. This leaves an opportunity for fraud on the side of a malicious online shopper. The fact that the e-Merchant cannot certainly authenticate a client, is enough for attempts at credit card fraud to be a persisting problem. Resulting statistics [PWC 98] show that credit verification systems are not advanced enough, resulting in false credit information being accepted as genuine. This inexorably hurts financially any e-Merchant having accepted fraudulent information as well as hurting unsuspecting people, whose credit information is in the possession of a malicious party.

2.4 Security Protocol Characteristics and Exploits

Current e-commerce trends [PWC 98] for securing Internet transactions reveal that the SSL protocol is seen and used by e-Merchants as the more secure alternative in providing a secure channel for transmission of sensitive information between online shoppers and electronic Merchants. The set of procedures provided by SSL allow for different options for securing and authenticating communicating parties [SSL 96]. There are three different options, which the protocol supports for the purpose of authentication:

- Anonymous communication; no authentication of any of the communicating parties.
- Server authentication; only the digital certificate of the server (e-Merchant) is transmitted to the client for authentication.
- Complete authentication; there is a mutual exchange of certificates between client and server.

The second and third option as listed above of the authentication process provide for a relatively sound structure for verification of e-Merchant (server) identification. The weakest option of the three listed is the anonymous connection between communicating parties, where no certificates are exchanged and thus no authentication is possible. This scenario is vulnerable to man in the middle attacks [SSL 96]. This can present a great cause for concern to any online shopper, as this weakness, if exploited properly can result in unsuspecting person's or entity's, credit information to be transmitted to a malicious third party, pretending to be a genuine e-Merchant [SON 97].

2.5 Secure Electronic Transaction (SET)

Based on the development of new technology such as smart cards, a new security protocols SET has emerged, whose purpose is to provide authentication and secure transactions between communicating parties [SET 97]. The main goal of SET is to allow specific cardholders and properly equipped web merchants to perform business transactions over an open and unprotected network. Such transactions, similar to many security payment protocols, are based on use of a set of cryptographic techniques, for the purpose of secure communication. The protocol further introduces a new approach to digital signatures, although it does not introduce any new algorithms or technologies. This approach sees the concept of dual signatures. This is done with the purpose of encapsulating an eventual payment to a merchant directly to the client's bank, as well with the purpose of creating an offer for goods or services to the merchant. If this offer is accepted, the merchant receives the full amount decided upon into his bank account, without being aware of the customers' credit particulars. In the same breath, the bank is not aware of the types of goods or services being purchased, or of their individual cost. This is all possible, with the existence of specific client and merchant side certificates. These are issued by each financial institution, which issues the credit smart card to clients and is in a relationship with the specific web merchant. The client certificate is stored on the client's smart card, but this certificate is optional and not compulsory. This coupled with the fact that not too many individuals are in the possession of a smart card reader, or in the case where they attempt to purchase goods or services online from a different form their own computer, this protocol, will not function properly in terms of authenticating the client as required by the protocol's functionality, thus presenting problems often encountered by web merchants. Such problems deal with trust in the funds and validity of the credit information provided, as well as the fact that the credit information may be valid, but stolen from its original owner.

2.6 Summary

This section presented certain security weaknesses and methodologies, which can be found in current e-commerce practices. The main concerns addressed here represent a low level of customer satisfaction of e-traders, based on poorly designed and implemented online trading practices, weak security measures for transmission of sensitive information, as well as lack of standardised practices for electronic transactions.

3. PROPOSED MODEL

3.1 Introduction

The proposed model outlines solutions for the problems encountered and described in the previous section as well as adding some extra features, which improve the overall security of the model. The model presents a methodology called Trusted Third Party (TTP), for securing a totally unsecured client, willing to perform online purchases, the authentication of communicating parties during this online transaction, as well as a secure transmission of sensitive information between them in the process of completing the online purchase.

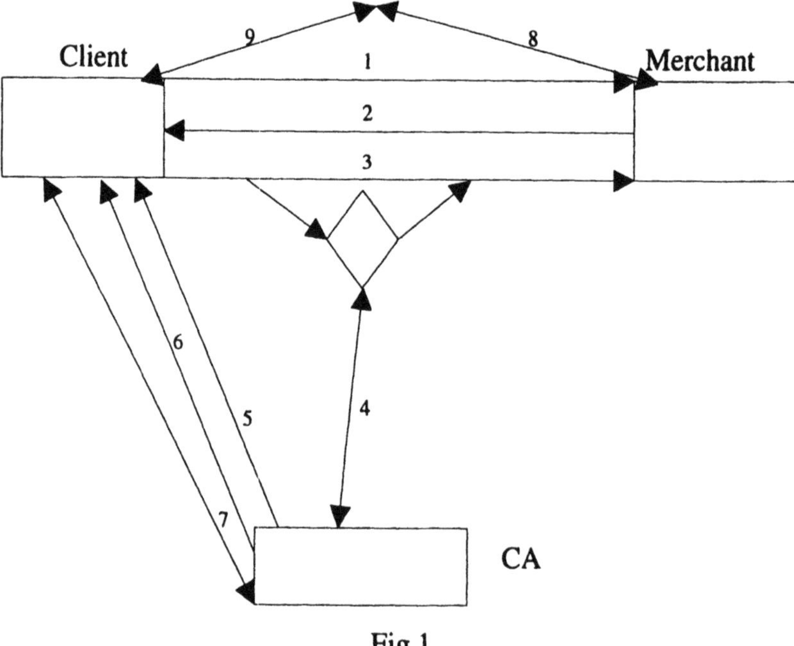

Fig 1.

3.2 Overview

The figure above describes the functionality of the Methodology presented here. An online transaction is usually initiated with an online shopper visiting the desired e-Merchant's web site (step 1). At this point the Merchant's server initiates a SSL session as specified in [SSL 96], with the online shopper. Once the initial SSL handshake procedure is initiated and the Server DC is delivered to the Client, the Server requests similar DC from the Client (step 2). If the Client is not in a possession of such certificate (step 3), the SSL session with the Client is interrupted and the client is notified that he/she needs to perform certain steps in order for the transaction to be secured. If he does decide to take up these steps, the shopper is redirected to the trusted CA's web site (step 4), while his session with the Merchant remains frozen. At this point the root certificate of the trusted CA is delivered to the shopper (step 5), followed by a small application, which is too installed at the client's machine (step 6). Immediately after that a java applet is delivered to the client (step 7), which communicates with the installed application from step 6 and generates two pairs of asymmetric keys, followed by the generation of corresponding DC. This completes the securing of the client and is followed by resumption of the frozen Merchant session. This sees a different java applet delivered to the client (step 8) used for credit information gathering and its encryption by the client residing application, as well as its transmission to the Merchant (step 9). Steps 8 an 9 do not follow through from entity to entity. This is done with the purpose of representing multiple transmissions of data between clients and merchant, once a secure communication between the two has been established and the appropriate authentication has been performed on either side.

3.2.1 Trusted third party

Based on the principal of trust, the trusted third party does not participate in any online transactions. Its sole purpose is to provide means of authentication and encryption for other entities, in order for them to be able to perform secure transactions over an unprotected network. Such attributes are provided by Certification Authorities [BPKIC 01]. The trusted third party within this methodology will be referred to as *Master CA*. The Master CA, consistent with the requirements of a CA, has a root certificate. One difference, which is vital to this section, is to mention that the root certificate of the Master CA is not self-signed, which is generally the practice of most well known CAs, it however is cross certified by a third party CA, which does not belong or is connected in any way to the Master CA.

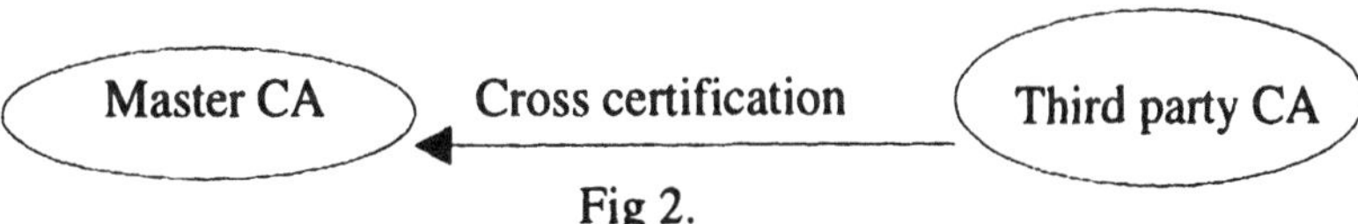

Fig 2.

3.2.2 E-Merchant

An e-Merchant is an online trader, providing sale of products or services. The Merchant requires payment in terms of a specific monetary currency, in return for his product or services. In most cases this payment is in the form of credit information, which is transmitted by the client to the Merchant. In order for the e-Merchant to be authenticated, he will need to have two Digital Certificates, holding the public keys for encryption and digital signature respectively. Intuitively, the private keys for those two corresponding public keys need to be in the possession of the Merchant and nobody else. The DCs are registered with the Master CA, with an appropriate chain of trust, and listed in this Master CA's Public Directory, if the Master CA itself has not issued them. The DCs need to be listed in the Master CA directory if not issued by the Mater CA, in order for the Master CA, serving as a point of trust, to be able to verify, the identity of the Merchant. The chain of trust to such a Digital Certificate needs to verifiable, in order for the Master CA to be able to trust its origin.

3.2.3 Online Shoppers (Clients)

These are people or entities, which wish to perform online transactions, in the form of purchases, from authentic e-Merchants. In order for an online shopper to be able to provide his or her sensitive credit information to the e-Merchant, he or she will require attributes similar to the Merchant's. These will be two pairs of public/private keys, for the purposes of digital signing of data and encryption respectively. The public key of each respective pair will need to be encapsulated in a Digital Certificate, which is either issued by the Master CA and thus signed by it, or is issued by any other CA with a verifiable chain of trust, and as with the e-Merchant scenario listed in the Master CA's Public Certificate Directory.

3.3 Initial Steps

The previous subsection described the minimum attributes required by two parties, in order for a secure communication to be established between them. The described scenario involved the introduction of a trusted third party, which does not take any part of any possible transactions involving an e-Merchant and an online shopper. Even though the online shopper and the e-Merchant are equipped with the necessary attributes to complete a secure online transaction, the two parties don't have a methodology in place, which will employ these attributes in a correct manner. Existing methodologies such as SSL have certain pitfalls, such as no online verification of DCs in

CRL, determining chain of trust of used certificates as well as guarantee of an existing standard for processing online transactions. Having said all that, most online shoppers are not equipped with any of the listed minimum attributes. Shoppers are purely restricted by the use of an Internet Browser (IB) and their concern of security of the transaction.

3.4 Obtaining the Master CA's Root Digital Certificate.

Once the online shopper is redirected to the Master CA's web site, the CA's Server detects his Internet Browser's make. That done, the shopper is further redirected (the whole process is automated) to download the Master CA's Root DC, which is cross certified by the maker of the shopper's Internet Browser. This done, the shopper's Internet Browser verifies the digital signature of the cross certifying third party CA (not the Master CA). This is possible, because each Internet Browser comes with the root certificate of the maker of the IB. This coupled with the fact that the root certificate of the Master CA is cross certified by the private key of the maker of the online shopper's IB makes this verification possible. From this point onwards the following procedures become more automated.

3.4.1 The Master CA's root certificate

Following standard asymmetric cryptography techniques, in order for a Digital Certificate to be generated there needs to be a public key of a public/private key pair encapsulated in it. The key pairs for the root certificate of the Master CA are generated using the standard RSA algorithm [PGP 95]. Use of other approved asymmetric algorithms can be equally as effective. The key length is of 2048 bits size. The private key of this pair is always kept with the Master CA. The public key is distributed to all known Internet Browser manufacturers, who based on it generate a Digital Certificate, which is signed with their own private key. Employing this technique the online shopper can be asserted that the received Master CA root certificate is indeed authentic and not fraudulent. Such approach can prove to be expensive, but it server to right purpose of secure transmission and identification of origin of transactions.

3.5 Background process

Once the Master CA's Root Certificate has been installed, any file or application signed with the private key of the Master CA will be guaranteed and be verifiable by the online shopper to be authentic and non malicious. This is used for the base of downloading a small application, which is installed and run on the client's machine. The application runs as a background process to the IB and is active throughout the whole time the client's machine is powered on. This application brings with itself the root certificates of the major Certification Authorities from around the world. These certificates are not hard wired into the application and are exchangeable, once they expire or become compromised. The application has networking capabilities, which will be discussed in the following sections. As part of the installation process, the application gives security permission to Java applets to interact with this background process. At no time however will an applet be able to control the background process.

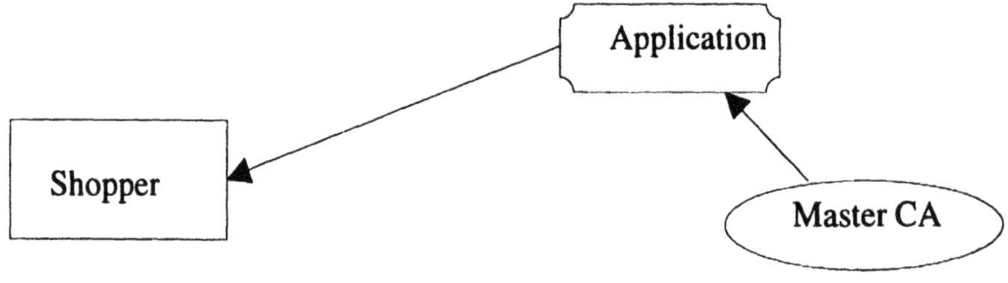

Fig 3.

The applet will simply be able to pass information to the process in the form of structured data.

3.6 Key pair generation

3.6.1 Review

The next step of the process sees the download of the application and its installation followed by continuation of the connection with the Master CA's server. After the installation procedure of the background process is complete, the client is redirected by the CA's server to download a Java applet.

3.6.2 Key generation and Digital Certificates

This applet is signed by the Master CA's private key. The purpose of this applet is to generate two pairs of keys using the RSA algorithm, or a similar asymmetric algorithm. These key pairs have the purpose of encryption and digital signing respectively. Once the applet is downloaded, its digital signature is verified by the IB. Following this, the two pairs of keys are generated. The public keys are passed to the background process, which signs them with the just generated signing private key, encrypts them with the public key of the Master CA, obtained from its certificate and passes them back to the applet. The applet sends this encrypted information back to the Master CA, which decodes this data and verifies the digital signature. Digital Certificates are created, encapsulating these public keys. The certificates are listed in the Master CA's Public Directory as well as these DCs being sent back to the applet, which together with the corresponding private keys are passed to the background process, for the purpose of storage and further use.

3.6.3 Communication between applet and background process

The communication between the applet and the background process is possible due to the fact that the applet has security permissions to communicate with this process. Before any communication between background process and applet is performed, the application verifies the digital signature of the applet, for reconfirmation of its origin. The communication between the background process and the applet is emphasized in figure 4.

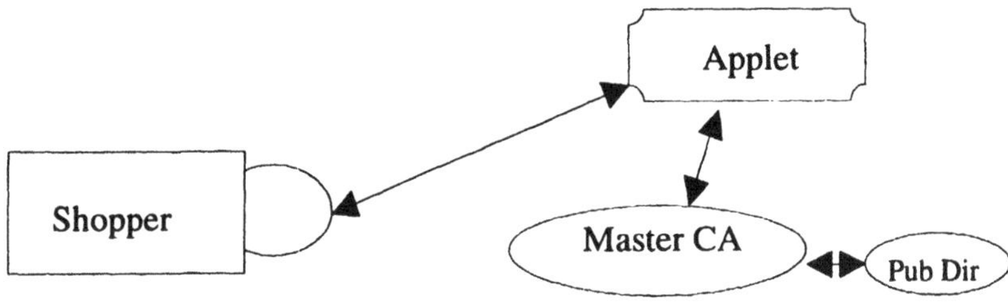

Fig 4.

3.6.4 Summary

This is the last step for securing a client in preparation for secure communication with a possible e-Merchant. This completes the process of establishing the basis of a methodology for secure and correct authentication of communicating parties, as well as for secure transmission of sensitive data over an open network. It is important to note that the process of securing the client, can be performed by anybody willing to adopt the methodology of secure communication as offered by the Master CA. This process does not have to be initiated by an e-Merchant who detects insufficient security on a client's machine; any concerned online shopper can initiate it.

3.7 Communication between shopper and e-Merchant

3.7.1 Download of Merchant applet

Once the securing of the client is complete, there is no need for the above-described procedures to be repeated ever again. The following step can be part of a resumption of a frozen session between a previously unsecured client with the e-Merchant, or as an initial step for submitting sensitive credit information by the client to the Merchant in completing the online transaction. This next step sees a Java applet downloaded from the e-Merchant's web site to the client's machine. The purpose of this applet is to collect sensitive credit information from the client and return it to the Merchant.

3.7.1.1 The Java applet

The Java applet is signed by the Master CA's private key. The applet takes as an external component the Merchant's Digital Certificate. The applet further has security permissions to communicate with the client's background process in the same manner described above as with the communication between the applet used for key pair generation by the Master CA.

3.7.2 Merchant Authentication

Before any sensitive information is entered by the online shopper in the downloaded applet, the IB first verifies the digital signature of the applet. Following this, the applet passes the Merchant's DC to the client's background process. This triggers an authentication procedure by the client's background process:

- Verification of the chain of trust of the Merchant's digital certificate.
- Online check of the Merchant's DC's ID in the issuer of this DC's CRL.
- Final online procedure, involving download of the Merchant's DC from the issuing CA's Public Directory. Then at the client's machine, a verification, of the main attributes of the DC, of the newly downloaded certificate versus the one received from the applet is performed.

In the possibility that at any authentication step yields a negative result, this would indicate an attempt for a security breach by a malicious party and the transaction is discontinued.

3.7.3 Information encryption

Once the authentication process on the client side is complete the shopper is prompted to enter credit information in the downloaded applet. This information is then passed to the client's background process, which includes the shopper's DC and encrypts the whole package with the Merchant's public key, signs it with his Digital Signature private key and passes it back to the applet which in turn transmits it to the online Merchant.

3.7.4 Client Authentication

Once the encrypted data is received, alongside with the Client's Digital Certificate, a chain of trust is established if possible, based on the existing trusted root certificates on the Merchant side. Following this, the certificate's authenticity is checked in the issuing CA's CRL as well as this certificate's validity is checked, by downloading this certificate from the CA's Public Directory and performing a comparison versus the certificate transmitted by the Client. If this authentication process does not run into any problems then credit card information, once decoded is verified using appropriate channels. This completes the transaction and the online shopper is notified of the fact that his/hers transaction has been performed or not.

3.7.5 Summary

The methodology outlined in this section (TTP), represents an effective process for secure authentication of two parties over an open network. The point of trust is a basic CA, which has established chain of trust. The structure of the methodology is such that it does not allow for anonymous communication between two parties, as complete mutual authentication is required before a transaction can be performed. The drawback of the proposed methodology is that it will affect the performance of any secure transaction between hosts and it will require a permanent connection to Cas, for the purpose of CRL verifications. The methodology avoids common pitfalls displayed by implemented technologies now in practice and thus is liable to raise consumer confidence in online trade.

4. EXTENDING THE MODEL

4.1 Weaknesses of proposed model

The above described model serves the purpose of securing a client with the required attributes for him or her to be able to perform a secure and authenticated transaction with another party, equipped with similar attributes, in a semi transparent manner. One of the attributes referred to is a pair of Digital Certificates. These certificates are issued to every client who has applied for them using the described above model or simply on his own initiative applied for them. These two DCs have the purpose of creating a digital identity for an applicant. This digital identity is created and based upon information provided by each applicant. This varies from first name and email address to surname and place of birth. Such information can be easily falsified and thus the digital identity based on it becomes untrustworthy. Examples of such digital identities, based on unverified information are represented by most level one certificates issued by most public CAs, to the general public.

4.2 Creating Trust

It becomes clear from the previous section that verification of user identity becomes vital to the proposed model's functionality. Such authentication of user identity can only be performed by the trusted third party and that is the *Master* CA. User authentication can be performed by a physical verification of the user details, into public records or relative government departments. This is assuming that the *Master* CA is based or has representation in each country, in which it has clients or applicants. Even if this was the case, physical verification of an applicant's identity would take a reasonable amount of time, far beyond what would be considered seamless and transparent process for securing a client, as specified by the proposed model. This would obviously not fit easily or at all in the described scenario of previous sections and would seriously impede the theoretical and practical flow of this methodology. Based on this, the need for an institution, which can easily and quickly verify the identity of an applicant, is required. Considering the fact that an applicant is at the point of purchasing goods or services, before he or she is redirected to the TTP, it must be apparent that this applicant is in the possession of some credit information such as a credit card, which is issued by a reputable financial institution, which are banks in most cases. Such institutions have performed a certain degree of identification of applicants, based on the fact that all applicants go through a thorough identification process before being issued with appropriate purchasing attributes, such as chequebooks or credit cards. This verification process is already performed at the point of a client wanting to purchase goods or services from and E-Merchant, as he or she is in the possession of at least a credit card.

The placement of trust in an issued digital identity in the form of a certificate, by the TTP, can be done by verifying that the person holding the credit-purchasing attribute is the same as the one to whom this credit attribute was issued. Such verification would confirm the identity of the applicant for a DC and thus place a great trust in the issued certificate. The trust in this certificate will be great because the process for issuing credit purchasing attributes such as credit cards, requires an extensive and thorough identification of the applicant, his or hers financial status as well as previous credit history. The above mentioned verification of credit attributes such as credit card versus the identification details of the applicant for a DC represent a simple mach of these two pieces of information in the financial issuing authority's database. This issuing authority could be represented by a bank but does not necessarily have to be.

4.2.1 Verification details

The specific details required for an applicant to be issued with a high trust certificate, deal with specific purchasing attributes e.g. credit card number, coupled with the card owner's name and a specific secret key. Such a key can be the pin number for this card or some secret code known only by the financial institution and the card owner. This would verify that the card is not merely stolen but it is the possession of its rightful owner. Such supply of information would be necessary for any other purchasing attribute apart from a credit card.

4.2.2 Finalization of authorization

Once all the required information is supplied to the TTP this data is passed to the relevant financial institution or issuing authority of purchasing attributes, such as credit cards, in order for this data to be verified. Once this information is verified, the financial institution can vouch for the identity of the applicant. This will place a very high trust in the resulting Digital Certificate(s). Following this approach, a financial institution such as a bank, or going higher up the hierarchy a credit card issuing enterprise such as Master Card or Visa, can serve as a authenticating parties in cooperation with a TTP in the process of creation and implementation of the proposed model.

5. CONCLUSION

This paper deals with presenting a new methodology for secure mutual authentication between two communicating parties over an open network. The process described here identifies and outlines clear steps for securing network communications and the data being transmitted in them. It takes a single point of trust in the name of a Certification Authority and with the help of small network based applications and applets constructs authentic way for identifying communicating hosts to one another via the use of asymmetric cryptography and Digital Signatures. The end result represents increased consumer confidence in a possible e-commerce environment, avoidance of current security pitfalls and potential decrease in credit card fraud.

6. LIST OF SOURCES CONSULTED

[SSL 96] The SSL Protocol Version 3.0, **November 18 1996** http://home.netsacape.com/eng/ssl3/draft302.txt

[PGP 95] The official PGP User's Guide, P.R. Zimmermann, **1995, MIT Press, USA**

[BPKIC 01] Basic Public-Key Infrastructure Characteristics, Marc Branchaud http://home.xcert.com/~marcnarc//PKI/thesis/characteristics **Jan 2001**

[DS 97] Decrypted Secrets Methods and Maxims of Cryptology, F.L. Bauer, **Springer-Verlag Berlin Heidelberg, 1997**

[SON 97] Security on the Net, Eddie Rabinovitch, http://www.cosmoc.org/ci/public/1997/mar/internet_column,html **1997**

[CTNS 00] The Concept of trust in Network Security, Entrust Technologies White Paper, **August 2000,** http://www.entrust.com

[VS 01] Outsourced Authentication Administrator's Guide, VeriSign, **January 2001.** http://www.verisign.com

[SET 97] Secure Electronic Transaction version 1.0, May 31, 1997, Master Card & Visa, http://www.visa.com/nt/ecomm/set/setprot.html

[PWC 98] Electronic Commerce/Internet Survey
http://www.pwcglobal.com/extweb/ncursvres.nsf

SOFTWARE-BASED RECEIPT-FREENESS IN ON-LINE ELECTIONS

Emmanouil Magkos*
Department of Informatics, University of Piraeus
80 Karaoli & Dimitriou, Piraeus, GREECE
emagos@unipi.gr

Vassilios Chrissikopoulos
Department of Archiving and Library Studies, Ionian University
Corfu, 49100, GREECE
vchris@ionio.gr

Nikos Alexandris
Department of Informatics, University of Piraeus
80 Karaoli & Dimitriou, Piraeus, GREECE
alexandr@unipi.gr

Abstract Electronic elections could be a viable alternative for real-life elections in a democratic society. In traditional elections, a voting booth does more than allow voters to keep their vote secret. The voting booth actually requires that voters vote secretly. If the privacy of the vote was allowed but not required, then a voter could easily sell his vote to a vote-buyer, or be coerced by a coercer. We present a receipt-free election scheme without making any hardware or physical assumptions about the communication channel between the voter and the voting authorities. Our solution is software-based i.e. voters are able to exercise their electoral rights from their home by using a personal computer with Internet access. The only physical assumption we make use of is an untappable channel between the two voting authorities that are employed in our scheme. This scheme satisfies most requirements of a secure electronic election. We make use of well-known cryptographic techniques such as time-lock puzzles and blind signatures.

Keywords: Receipt-freeness, electronic voting, privacy.

*Research supported by the Secretariat for Research and Technology of Greece.

1. Introduction

Due to the rapid growth of the Internet, electronic voting could be a viable alternative for governmental elections, especially in the case of geographically distributed voters with access to open computer networks. If not carefully designed and implemented, e-voting systems can be easily manipulated, thus corrupting election results or violating voters' privacy.

In traditional elections, a voting booth does more than allow voters to keep their vote secret. The voting booth actually requires that voters vote secretly. If the privacy of the vote was allowed but not required, then a voter could easily sell her vote to a vote-buyer, or be coerced by a coercer. All receipt-free schemes met in the literature use hardware assumptions to achieve receipt-freeness. In [15] there are tamper-resistant smartcards that keep some information secret from the voter. Most other schemes [1, 2, 9, 10, 12, 13, 14, 19] make physical assumptions about the communication channel between the voter and the election authorities. More specifically, they assume the existence of :

- *Untappable channels* from the voter to the authority [13, 14]
- *Untappable channels* from the authority to the voter [1, 9, 10, 19]
- *Physical Voting Booths* [2, 12].

In [9], it is argued that "one-way channels from the authorities to the voters are the weakest physical assumption for which receipt-free voting protocols are known to exist". We believe that these physical assumptions are unsatisfactory: If the underlying communication structure consists of untappable channels between the voting authority and secure dedicated machines (where voters vote), then there is no point of quitting the traditional elections. Real life citizens in a democratic society, who find it inconvenient to go to the polls (and so they finally abstain from the elections) will find it equally inconvenient to cast their vote from a physical voting booth in a dedicated computer network. Note that untappable channels will also force the voter to use specified voting locations.

Our Contribution. We present a software-based receipt-free election scheme, which is secure against a coercer who has tapped all the communication lines between a voter, say Victor, and the voting authorities. Victor's vote is a computational *time-lock puzzle* [17], i.e., it requires a precise amount of time (real time, not CPU time) to be solved, unless a trapdoor information is known in advance. In our election scheme, the trapdoor information is only known to a voting authority. A second authority exists to make sure that votes remain secret until the end of the voting period. A coercer, who wants to find out who Victor voted for, has no other way than running a dedicated computer continuously for a certain amount of time. Even if Victor has incentives to prove his vote to a vote-buyer, there are no means to prove it, since he does not know the trapdoor information. We do not assume any untappable channels between Victor and the voting authority, or any hardware devices. The only physical assumption we make use of is an untappable channel between the two voting authorities that are employed in our system.

The cost paid for receipt-freeness is that the voter constructs her vote inefficiently, by repeatedly squaring a given value, for a significantly large amount of time. However, we believe that this is a minimal tradeoff for a software-based receipt-free solution. To our knowledge, our scheme is the only receipt-free scheme in the literature without the physical assumption of an "untappable" channel between the voter and the voting authority. Voters are able to exercise their electoral rights from their home by using a personal computer with Internet access. Furthermore, our scheme satisfies most security requirements met in the literature.

2. A Model for Software-based Receipt-Freeness

In our model we assume that a coercer may have tapped the communication channel between the voter and the voting authority. It is clear enough that the vote should be encrypted, for vote secrecy. The *trapdoor* information for the encrypted vote may consist of a secret decryption key and/or the randomness used in a probabilistic encryption scheme. If this trapdoor is in the possession of the voter (e.g. as in [3, 13, 14]) then it could also serve as a receipt for the vote. Even if the voter "lies" about the encrypted vote [1, 9, 10, 19], a coercer who taps the communication channel will eventually find out the value of the vote by eavesdropping on the confidential information exchanged between the voter and the authority. Note that simple encryption does not serve our purposes: a coercer will tap the encrypted message as it is being sent from the voter to the authority (or vice versa), and then require the voter to reveal the trapdoor information. Even worse, the coercer may demand that the voter uses some specific randomness. To summarize: the simplest bit of information that will make the voter's life easier during the construction of the encrypted vote, may also make the coercer's life easier. Thus, software-based receipt-freeness in the presence of a coercer who taps communication lines can *only* be achieved if the voter does not use any secret information other than the vote itself.

We came up with a variation of the *time-lock* concept, as has been described by Rivest, Shamir and Wagner [17]. The idea is based on *preprocessing*: the voter selects a vote from a set of valid votes and constructs a time-lock puzzle of the vote by repetitively squaring a specific value (which is not secret). The number of squarings is also a public parameter. In [17] the user does this efficiently because she knows some trapdoor information. In our model, the user does not know the trapdoor information. The trapdoor information is possessed by a voting authority and it will be used at the end of the voting phase to reveal the cleartext vote. Thus, the voter constructs his vote *inefficiently* by executing an "intrinsically sequential" process.

A coercer, who taps the communication line between the voter and the voting authority, will get the time-lock puzzle of the vote, as it is being delivered to the authority. Even with the help of the voter, there is no way to reverse the time-lock process: the voter does not know the trapdoor information, so the coercer will have to run a dedicated computer for a specific amount of time. This time can be determined by an independent authority who sets the public parameters, e.g. the number of squarings for each puzzle, so as to prevent massive coercion in a large-scale election: assuming that each voter performs n squarings, a coercer will have to perform nk squarings to coerce k voters. However the voter too constructs his vote inefficiently, but we believe that this is a minimal tradeoff for a software-based receipt-free solution that does not employ untappable channels between the voter and the voting authority.

The only physical assumption we make use of is an untappable channel between the two voting authorities employed in our system. This is acceptable, since our main goal was to abolish the necessity of a physically secure channel between the voter and the authority. We believe that an untappable channel between two authorities that belong to a distributed set of voting authorities, is a minimal physical assumption for a receipt-free scheme. We could remove this physical assumption by requiring that there is only one voting authority, but in that case, and unless full trust was granted to this authority, fairness would have been sacrificed: if the authority possesses the trapdoor information, then votes may be revealed before the end of the voting period.

3. Building Blocks

Our voting scheme makes use of *blind signatures* [4], which is a well known technique, already implemented with the RSA algorithm [16, 20]. Blind signatures are the equivalent of signing carbon-paper envelopes: a user seals a slip of a paper inside such an envelope, which is later signed on the outside. When the envelope is opened, the slip bears the carbon image of the signature. Furthermore, users in our scheme lock their votes in a *time-lock puzzle.* The mechanism is a variation of a well-known technique [17] and is presented below.

3.1. Time-lock Puzzles

Suppose that Alice wants to encrypt a message M so that Bob can decrypt it after a period of T seconds. T is a real (not CPU) time period, given that Alice knows (or approximately assumes) in advance the CPU power of Bob. In [17], Alice generates a composite modulus $n = pq$ as the product of two large primes p and q. Then, Alice computes $\Phi(n) = (p-1)(q-1)$ and $t = TS$, where S is the number of squarings modulo n per second that Bob can perform. Alice chooses a key K for a symmetric cryptosystem and encrypts M with key K, thus getting $C_M = Enc(K, M)$. In order to hide K, she picks a random a modulo n and encrypts K as:

$$C_K = K + a^{2^t} (mod\ n) \tag{1}$$

To do this efficiently, Alice uses the trapdoor information $\Phi(n)$ that only she knows: She first computes $e = 2^t (mod\ \Phi(n))$ and then $b = a^e (mod\ n)$. The public output of the puzzle is the set (n, a, t, C_M, C_K). Since Bob does not know the factors p and q, computing $\Phi(n)$ from n is provably as hard as factoring n. Bob has no way of computing a^{2^t}, other than starting with a and perform t sequential squarings, each time squaring the previous result. The computational problem of performing these squarings is not parallelizable: having two computers is not better than having one computer.

Our variation. In our model, Alice is the voter and Bob is the coercer. Alice *does not know* the trapdoor information $\Phi(n)$ (if she knew it she could hand it over to Bob, e.g. in a vote-selling scenario), so she cannot construct the puzzle efficiently. In addition, there are two voting authorities. The first authority selects n, p and q, and publishes n and t, where t is the number of squarings that Alice has to perform. Alice selects a as previously and computes a^{2^t}. Alice's vote v takes the place of the key K in equation (1), thus yielding:

$$C_V = v + a^{2^t} (mod\ n) \tag{2}$$

The public information will now be the set (n, a, t, Cv). When the time comes, Alice uses a clear channel to submit the time-lock puzzle of her vote to the second voting authority. The first voting authority, who possesses the trapdoor information $\Phi(n)$, will later cooperate with the second authority to decrypt the submitted votes. In Section 4, our voting protocol is presented in detail.

An efficient solution with "secure" hardware. Another solution would be each voter to be equipped with a tamper-resistant smartcard. During an off-line registration protocol, this smartcard would be provided with the trapdoor information $\Phi(n)$. Later, during voting phase, the voter would provide the smartcard with her preferable vote, and the smartcard would use the trapdoor information $\Phi(n)$ to construct the time-lock puzzle in an efficient way. To reveal the vote, the coercer would either have to tamper with the smartcard or solve the time-lock puzzle.

4. A Receipt-free E-voting Scheme

In our protocol there are N voters and two authorities, the Registrar and the Voting Center. The Registrar acts as an intermediate between the voter and the Voting Center, while the Voting Center is responsible for tallying the votes. We assume that each authority is *semi-trusted* [7], i.e., the authority may misbehave but will not conspire with another party. We also make use of a bulletin board, which is publicly readable. Only the Voting Center can write to it, and nobody can delete from it. The Voting Center is committed to everything that is published on the bulletin board.

There is a certificate infrastructure and all participants are legally bound by their signatures. Voters and authorities possess a private/public key pair for signature and encryption as well as the corresponding certificates, issued by a trusted Certification Authority. We also assume that there is an untappable channel between the Registrar and the Voting Center. Communication between voters and authorities takes place through an *anonymous channel*: voters can send/accept messages that cannot be traced (e.g., by using traffic analysis). For example, e-mail anonymity can be established with Mixmaster re-mailers [5], and HTTP anonymity can be established with services such as the Onion-Routing system [8]. The election protocol is depicted on Figure 1. It is split into four stages, the *Authorizing* stage (Steps 1-2), the *Voting* stage (Steps 3-5), the *Claiming* stage (Step 6) and the *Tallying* stage (Steps 7-8).

Authorizing Stage. A voter, say Victor, wishes to get a certified pseudonym that will identify him to the Voting Center. Victor creates a private/public key pair (SK_{PS}, PK_{PS}), blinds PK_{PS} (the public tallying key) to create the blinding b_1 and then signs a message consisting of b_1 and the unique election identification number $Elect_{id}$ (Step 1).

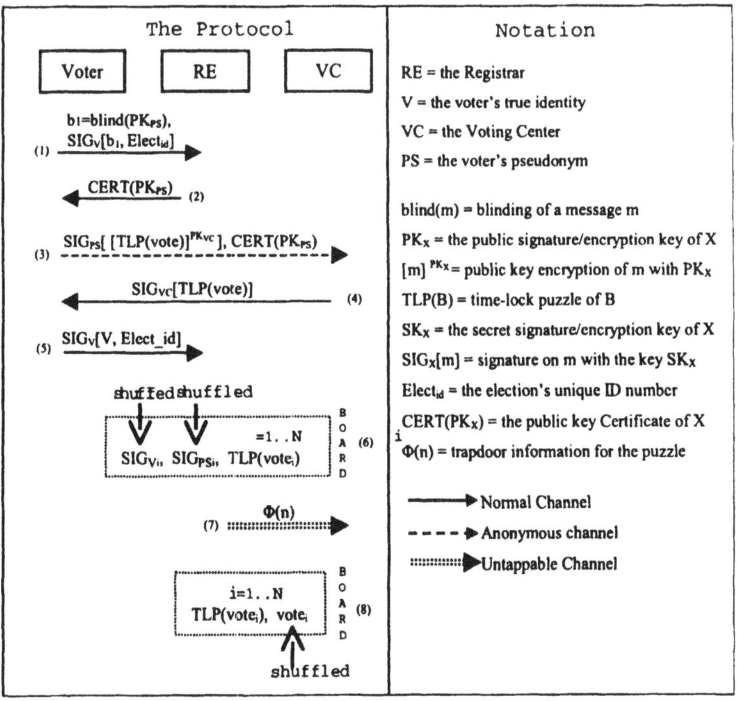

Figure 1. A receipt-free election scheme (software-based)

Victor sends these to the Registrar and gets the blinding signed by the Registrar (Step 2). Victor unblinds the Registrar's signature on b_1 and is left with a certificate of the public tallying key, $CERT(PK_{PS})$. This certificate will be used later by the Voting Center to verify signatures that are made with the secret tallying key, SK_{PS}. The public key PK_{PS} will be Victor's official pseudonym.

Voting Stage. In Step 3, Victor, who has already constructed a time-lock puzzle of his vote, $TLP(vote)$, encrypts it with the public key of the Voting Center, and signs the result using his secret tallying key, thus producing $SIG_{PS}[[TLP(vote)]^{PK_{VC}}]$. He anonymously sends this to the Voting Center, along with the certificate of his public tallying key. The Voting Center verifies the signature, decrypts the message, stores the time-lock puzzle in a local database and returns, in Step 4, a signature on the puzzle, $SIG_{VC}[TLP(vote)]$. This can be seen as a receipt that the Voting Center has accepted the time-lock puzzle of the vote. At some time later, in Step 5, Victor uses his authentic signature key to sign a message consisting of his true identity V and the $Elect_{id}$ number. He then sends the signature to the Voting Center.

Claiming Stage. In step 6, the Voting Center publishes on the board, in random order, the list of the authentic and pseudonymous signatures SIG_{V_i} and SIG_{PS_i}, $i = 1, ..., N$. The Voting Center also publishes all the time-lock puzzles $TLP(vote_i)$ of the votes that have been successfully submitted. In case Victor's time-lock puzzle is not published, he can protest by broadcasting $SIG_{VC}[TLP(vote)]$, with no need to reveal in which way he actually voted. This is called an "open objection to the tally", introduced by Sako in [18].

Tallying Stage. In Step 7, the Registrar sends the secret trapdoor $\Phi(n)$ to the Voting Center, by using an untappable channel. No one, except the Voting Center, can have access to $\Phi(n)$. The Voting Center uses $\Phi(n)$ to solve the time-lock puzzles of the votes. In Step 8, the Voting Center publishes in clear the results of the election, i.e. the list of the votes $vote_i$, $i = 1, ..., N$. The Voting Center also publishes a list with the corresponding time-lock puzzles of the votes, $TLP(vote_i) = [TLP(vote_1), ..., TLP(vote_n)]$.

5. Security Analysis

We evaluate the security of our scheme by examining some basic requirements, which most researchers seem to agree upon [6, 20]:

Eligibility. (Only authorized voters are able to vote). In Step 1, Victor signs a message using his authentic signature key. The Registrar checks the eligibility of each user who submits a tallying key for certification.

Unduplicability. (No one is able to vote more than once). The Registrar will not issue more than one tallying keys for each voter. In Step 6, all the authentic signatures of the voters are published. Consequently, it is not possible to exist more tallying keys than authentic public keys, so the Registrar cannot misbehave without being caught.

Untraceability. (All votes remain anonymous). When Victor submits a tallying key for certification, he signs a message and the Registrar checks his identity. However, the tallying key is blindly signed by the Registrar in Steps 1-2. Consequently, the Registrar cannot trace any signature SIG_{PS_i} published in Step 6, back to Victor's real identity. Furthermore, Victor in Step 3 uses an anonymous channel to submit his validated time-lock puzzle. The puzzle cannot be traced back to its sender, since it is signed under a certified pseudonym (the tallying key).

The link between Victor's pseudonym and his real identity cannot be done by either authority.

Fairness. (All ballots remain secret while voting is not completed). The trapdoor information necessary to solve the puzzle, is in the possession of the Registrar. Victor encrypts the time-lock puzzle of his vote with the public encryption key of the Voting Center, and sends it to the Voting Center. The Voting Center will not publish the time-lock puzzles until the end of the voting period. Fairness is achieved, as long as the Registrar and the Voting Center do not combine their knowledge. Neither the Registrar nor the Voting Center can break fairness by themselves. Since the Registrar and the Voting Center are assumed to be semi-trusted, this requirement is satisfied.

Accuracy. (No one is able to alter/delete anyone else's vote). In Step 6, the Voting Center commits to the time-lock puzzles of all the votes and cannot alter them, according to the properties of the bulletin board. Every voter, whose time-lock puzzle has not been taken into account, can make an "open objection to the tally".

Atomic Verifiability. (Voters are able to verify that their vote has been counted correctly). In Step 6, all the time-lock puzzles of the votes are published by the Voting Center. Victor can check that his time-lock puzzle has been published on the board. If not, Victor makes an open objection: he anonymously broadcasts the receipt that was sent to him in Step 4.

Receipt-Freeness. (No voter is able to prove the value of its vote). The receipt freeness property is separately discussed in Section 2. It must be noted that the scenario of a coercer who observes the voters at the moment they vote, is not addressed at all. This attack cannot be prevented by any e-voting scheme and is rather unrealistic in large-scale elections.

Responsibility. (Eligible voters who have not voted can be identified). This is an optional requirement, desirable in Australian elections [11]. All voters, who receive in Step 4, an acknowledgment of their votes from the Voting Center, sign a message by using their authentic signature keys and send this message to the Registrar, in Step 5. The Registrar has already received, in Step 1, the authentic signatures of all eligible voters, so he is able to identify, by comparing the corresponding lists, the eligible voters who have not voted.

6. Discussion

We have presented a receipt-free election scheme, which satisfies most requirements of a secure election. We do not assume any hardware devices or untappable channels between the voter and the voting authorities. We make use of well-known cryptographic primitives that have been implemented. Time-lock puzzles, while being very difficult in their solution, are quite efficient in their construction. The problem with our scheme is that we sacrifice efficiency in order to achieve software-based receipt-freeness. While the computations *during* the election are done quickly and in few steps, the computations made by the voter *before* the election (the preprocessing for the time-lock puzzle) are not done in a reasonable amount of time. This time is determined by an authority, and has to be long enough to discourage massive coercion of voters. Yet, as noted in Section 3.1, our scheme could be relaxed to become an efficient scheme with smartcards. In such case, however, the scheme would be a hardware-based solution.

References

[1] D. Alpert, D. Ellard, O. Kavazovic, M. Scheff. *Receipt-Free Secure Elections 6.857 Final Project*, 6.857 Network and Computer Security, 1998, http://www.eecs.harvard.edu/~ellard/6.857/final.ps.

[2] J. Benaloh, D. Tuinstra. *Receipt-free secret-ballot elections*, 26th Annual ACM Symposium on the Theory of Computing, Proceedings, 1994, pp. 544-553.

[3] R. Canetti, C. Dwork, M. Naor, R. Ostrovsky. *Deniable Encryption*, Advances in Cryptology - CRYPTO 97, Proceedings, Lecture Notes in Computer Science, Vol. 1294, Springer-Verlag 1997, pp. 90-104.

[4] D. Chaum. *Blind Signatures for Untraceable Payments*, Advances in Cryptology - CRYPTO 82, Proceedings, Plenum Press 1982, pp. 199-203.

[5] L. Cottrell. *Mixmaster and Remailer Attacks*, http://obscura.obscura.com/~loki/remailer/remailer-essay.html.

[6] L. Cranor, R. Cytron. *Sensus: A security-conscious electronic polling system for the Internet*, Hawaii International Conference on System Sciences, Proceedings, 1997, http://www.research.att.com/~lorrie/pubs/hicss/hicss.html.

[7] M. Franklin, M. Reiter. *Fair exchange with a semi-trusted third party*, 4th ACM Conference on Computer and Communications Security, Proceedings, ACM 1997, pp. 1-6.

[8] D. Goldschlag, M. Reed, P. Syverson. *Onion Routing for Anonymous and Private Communications*, Communications of the ACM, Vol. 42(2), pp. ACM 1999, pp. 39-41.

[9] M. Hirt, K. Sako. *Efficient Receipt-Free Voting Based on Homomorphic Encryption*, Advances in Cryptology - EUROCRYPT 2000, Proceedings, Lecture Notes in Computer Science, Vol. 1807, Springer-Verlag 2000, pp 539-556.

[10] B. Lee, K. Kim. *Receipt-free Electronic Voting through Collaboration of Voter and Honest Verifier*, JWISC 2000, Proceedings, 2000, pp. 101-108.

[11] Y. Mu, V. Varadharajan. *Anonymous Secure e-voting over a network*, 14th Annual Computer Security Application Conference, Proceedings, IEEE Computer Society 1998, pp. 293-299.

[12] V. Niemi, A. Renvall. *How to prevent Buying of Votes in Computer Elections*, Advances in Cryptology - ASIACRYPT 94, Proceedings, Lecture Notes in Computer Science, Vol. 917, Springer-Verlag 1994, pp. 141-148.

[13] T. Okamoto. *An Electronic Voting Scheme*, IFIP '96, Proceedings, Advanced IT Tools, Chapman & Hall 1996, pp. 21-30.

[14] T. Okamoto. *Receipt-Free Electronic Voting schemes for Large Scale Elections*, Workshop of Security Protocols '97, Proceedings, Lecture Notes in Computer Science, Vol. 1163, Springer-Verlag 1996, pp. 125-132.

[15] A. Riera, J. Borrell, J. Rifa. *An uncoercible verifiable electronic voting protocol*, 14th International Information Security Conference IFIP/SEC'98, Proceedings, 1998, pp. 206-215.

[16] R. Rivest, A. Shamir. *A Method for Obtaining Digital Signatures and Public-Key Cryptosystems*, Communications of the ACM, Vol. 21, ACM 1978, pp. 120-126.

[17] R. Rivest, A. Shamir, D. Wagner. *Time-Lock Puzzles and Timed-Released Crypto*, LCS Technical Memo MIT/LCS/TR-684, 1996, http://www.theory.lcs.mit.edu/~rivest/RivestShamirWagner-timelock.ps

[18] K. Sako. *Electronic Voting Scheme Allowing Open Objection to the Tally*, IEICE Transactions on Fundamentals of Electronics, Communications and Computer Sciences, Proceedings, Vol. E77-A(1), 1994, pp. 24-30.

[19] K. Sako, J. Killian. *Receipt-Free Mix-Type Voting Schemes - A practical solution to the implementation of voting booth*, Advances in Cryptology - EUROCRYPT 95, Lecture Notes in Computer Science, Vol. 921, Springer-Verlag 1995, pp. 393-403.

[20] B. Schneier. *Applied Cryptography - Protocols, Algorithms and Source Code in C*, 2nd Edition, 1996.

ID-BASED STRUCTURED MULTISIGNATURE SCHEMES

Chih-Yin Lin, Tzong-Chen Wu* and Jing-Jang Hwang
Institute of Information Management, National Chiao Tung University, Hsinchu, Taiwan, R.O.C.
**Department of Information Management, National Taiwan University of Science and Technology, Taipei, Taiwan, R.O.C.*

Abstract: The signing structure of a multisignature scheme specifies the signing order for all signers when signing messages, and any multisignature not obeying the specified signing order will be verified as invalid. In accordance with the different responsibilities of the participant signers, the signing structure of a multisignature scheme could be further classified as the following three types: serial, parallel and mixed, where the mixed structure is regarded as the mix of the serial and the parallel. Based on the well-known ID-based public key system, we will propose three ID-based structured multisignature schemes and each scheme respectively realizes the serial, parallel and mixed signing structures. In the proposed schemes, the length of a multisignature is fixed and the verification of a multisignature is efficient, regardless of the number of signers involved. Besides, any invalid partial multisignature can be effectively identified during the generation of the multisignature.

Keywords: Multisignature, structured multisignature, ID-based public key, signing structure.

1. INTRODUCTION

A multisignature scheme allows multiple signers to sign messages in which all signers have to sign and individual signer's identity can be identified from the multisignature [2-6, 8-10, 13-16, 19]. Furthermore, a structured multisignature scheme [4, 6] is a multisignature scheme that additionally requires all signers to obey a predefined signing structure when signing messages, and any multisignature generated without obeying the specified signing structure will be verified as invalid. The signing structure of a multisignature scheme indicates the signing order among all participant signers when signing messages, As a consequence, the multisignature of a message in a structured multisignature scheme is said to be valid when the following conditions are satisfied: (i) All signers had signed the message; (ii) All signers perform their signing operations in compliance with the specified signing structure; (iii) The multisignature and all partial multisignatures generated during the multisignature generation process have been successfully verified. Typical applications of the structured multisignature scheme are multisignatures used in a corporate organization or hierarchical environment. For example, a legitimate working report should be signed accordingly in the order of the operators, the section leader and the department manager. Signing structures can be classified into three basic types: serial, parallel, and mixed, where the mixed structure is the mix of the serial and the parallel. For the serial structure, all signers sign messages in a predetermined sequence, and hence the generated multisignatures are sensitive to the signing order. As to the parallel structure, all signers sign messages in a parallel manner and the generated multisignatures are independent of the signing order. In the mixed structure, the signing structure is composed by substructures that could be serial, parallel, or another mixed structure, and the generated multisignatures are sensitive to the signing order specified in the corresponding signing structure. Figure 1 depicts these three types of signing structures.

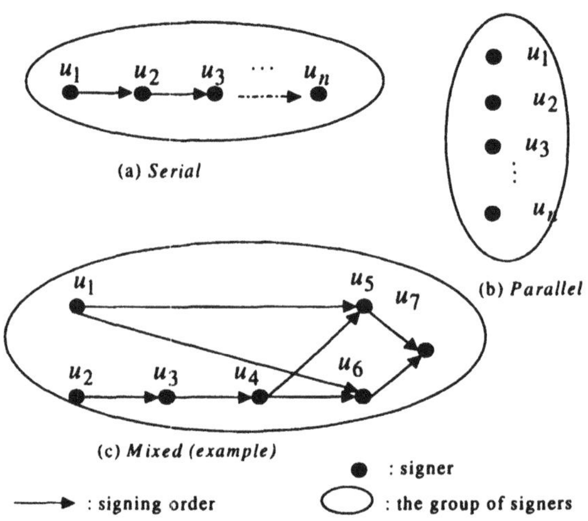

Figure 1 – Three types of the signing structure.

Most of the previously proposed multisignature schemes are irrelevant to signers' signing order, while some others are order-sensitive. The schemes presented in References 2, 3, 5, 8, 13, 14, and 19 are order-irrelevant, and the schemes presented in References 4, 6, 9, 10, 15, and 16 are order-sensitive. Among the order-sensitive schemes, the schemes proposed by Harn and Kielser [9], Itakula and Nakamura [10], and Okamoto [16] are RSA-like multisignature schemes in which the signers' signing order has to be properly arranged by different modules of their public keys; otherwise, messages to be signed might be modularly truncated. Besides, the length of the multisignature and the verification time required by these RSA-like schemes varies proportionally with the amount of the signers participated. In 1998, Doi *et al.* [6] firstly proposed a multisignature scheme considering the mixed signing structure. They used structured group identity and proposed two structured multisignature schemes for common modular RSA-type and ElGamal-type signature schemes. However, the length of the multisignature generated by their schemes varies with the number of the signers involved. Later, Burmester *et al.* [4] proposed an ElGamal-type multisignature scheme with a structured public key approach. In their scheme, the secret and the public keys for each signer could be generated either by a trusted centre or by cooperative signers using a distributed protocol. Moreover, Burmester *et al.* assumed that there exists at least one honest signer for their scheme to be secure. This assumption is somewhat less practical and incompatible especially when applying to a delegation scheme, i.e. proxy signature [11-12], in which the original signer has to consider the threat that all (proxy) signers in the signing structure may commit frauds or collusions.

Based on the well-known ID-based public key systems [7, 18], we propose three structured multisignature schemes whose security is based on the difficulty of solving discrete logarithm modulo a large composite (DLMC) [1] and factorising a large composite (FAC) [1, 17]. Since ID-based digital signature and multisignature schemes [7, 18-19] use the identity of the signer as the public key, our scheme has the advantage that the signature verification requires no extra interaction for public key verification. The proposed schemes have the following merits:

(1) The length of the multisignature is fixed to different messages.

(2) The length of the multisignature is fixed regardless of the number of signers.

(3) The computation cost required for the multisignature is efficiently fixed to the amount of signers participated.

(4) Any violation to the signing order will be detected and identified immediately.

This paper is sketched as follows. After the introduction, we specify the notations, parameters and signing structures in section 2. In section 3, we will propose the serial, the parallel and the mixed structure multisignature schemes. Security analysis is given in section 4. We conclude the paper in section 5.

2. NOTATIONS AND PARAMETERS

Let $G=\{u_1,u_2,...,u_n\}$ be a group consists of n signers and $sk_i=(sk1_i,sk2_i)$ be u_i's private key. The verification key for the partial multisignatures generated by u_i is named "partial verification key" and denoted by PK_i. The verification key for verifying the multisignatures generated by G is named "public verification key" and denoted by VK_G.

For each signer $u_i \in G$, his identity ID_i is a message digest of his/her public identification information I_i using a one-way hash function, said F, such that $ID_i=F(I_i)$. As defined in other ID-based crypto schemes [7, 18-19], I_i can be a combination of u_i's name, age, gender, telephone number or home address, provided that this combination can uniquely identify u_i. Note that a system authority SA is assumed [7, 18-19] for setting up the ID-based cryptosystem.

2.1 Signing Structure

Two types of notations are used for describing the signing structures. SER[ ] denotes the serial structure; and PAR[ ] the parallel structure. For $G_1=\{u_1,u_2,u_3\}$, if the legal signing sequence is $<u_1,u_2,u_3>$, then the corresponding signing structure is SER$[u_1,u_2,u_3]$. Another example is, for $G_2=\{u_1,u_2,u_3,u_4\}$ with a mixed signing structure SER[u_1, PAR[u_2, u_3],u_4], there are exactly two legal signing sequences, which are $<u_1,u_2,u_3,u_4>$ and $<u_1,u_3,u_2,u_4>$. Furthermore, we can use a diagram to represent the corresponding signing structure as in Figure 1 and 2. In the diagram, each node indicates a signer and each arrow implies the signing order for the two signers it connects. If an arrow points from u_i to u_j, it means u_j should sign after u_i signs. In the above example group of signers G_2, it can be draw as in Figure 2(a). Notably, in order to facilitate the tasks performed in the structured multisignature schemes described later, we add two dummy nodes s and t to the diagram representation where s and t denote the start node and terminate node, as shown in Figure 2(b). The general diagrams for a group of serial signers, a group of parallel signers and an example diagram for a group of mixed-structure signers are shown (i.e. SER[PAR[u_1,SER[u_2,u_3,u_4]],PAR[u_5, u_6],u_7]) respectively in Figure 1.

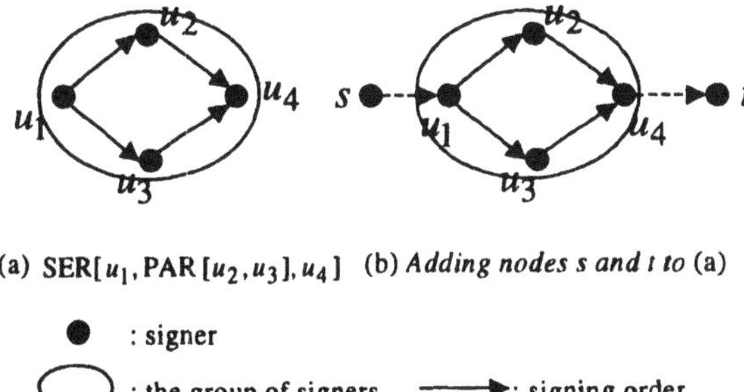

Figure 2 – The diagrams of an example signing structure.

2.2 System Parameters

SA initialises the ID-based public key system applicable for structured multisignatures by first preparing the following parameters.

p, q : two large prime integers, where $2p+1$ and $2q+1$ are also primes.
N : the product of $2p+1$ and $2q+1$ that $N=(2p+1)\cdot(2q+1)$.
ω : the product of p and q that $\omega = p\cdot q$.
α : a base of order ω modulo N.
r : a random number, where $r \in Z^*_\omega$.
β : $\beta = \alpha^r \bmod N$.
f, h : two hash functions, where $f(x) < \min(p,q)$ and $h(x) < \min(p,q)$.

SA keeps p, q, ω and r secret, while publishing N, α, β, f and h. Note that f is used to generate the public identities and verification keys and h is used to produce the message digest of the message to be signed. Throughout this paper, x^{-1} denotes the inverse of x modulo ω.

2.3 Public Verification Keys

SA generates VK_G for G and PK_i for each $u_i \in G$ by the following rules. For serial structure, signers' public identities, i.e. ID_i's, are concatenated, and for parallel structure, signers' public identities are first sorted then concatenated, to be the input of the function *f*. The output of *f* is the value of the public verification key.

Notice that the reason why we sort and concatenate the identities of signers in the parallel structure is to provide uniqueness of the verification key. Consequently, the possibility of the existence of two identical verification keys can be eliminated. To achieve this, we can use a function, said $\zeta(\ )$, that takes a variant numbers of values as input, sorts the input values, and finally outputs the value of the concatenation of the sorted input values. For example, the output of $\zeta(2983,9213,7615,1003,8714)$ will be 10032983761987149213. For $G=\{u_1,u_2,u_3,u_4\}$ and its signing structure $\mathrm{SER}[u_1,\mathrm{PAR}[u_2,u_3],u_4]$, $VK_G = f(ID_1 \,\|\, \zeta(ID_2,ID_3) \,\|\, ID_4)$, $PK_1 = f(ID_1)$, $PK_2 = f(ID_1 \,\|\, ID_2)$, $PK_3 = f(ID_1 \,\|\, ID_3)$, and $PK_4 = VK_G = f(ID_1 \,\|\, \zeta(ID_2,ID_3) \,\|\, ID_4)$.

3. THE PROPOSED MULTISIGNATURE SCHEMES

The multisignature schemes for serial, parallel and mixed signing structures are presented respectively. Each proposed scheme consists of three phases: key generation, multisignature generation and multisignature verification. In key generation phase, the system authority generates the private key for each signer. In the multisignature generation phase, each signer follows the signing structure to sign messages after verifying the partial multisignatures generated by the preceding signers. Finally in the multisignature verification phase, the verifier verifies the validity of the multisignature. Details are given in the followings.

3.1 For serial signing structure

Without loss of generality, assume the group of signers $G=\{u_1,u_2,\ldots,u_n\}$ is associated with the signing structure $\mathrm{SER}[u_1,u_2,\ldots,u_n]$. That is, all $u_i \in G$ have to sign messages by following the serial order $u_1,u_2,\ldots,u_n$ for generating a valid multisignature. The scheme is stated as follows.

Key generation phase:

SA prepares the partial verification keys and public verification key as $PK_i = f(ID_1 \,\|\, ID_2 \,\|\, \ldots \,\|\, ID_i)$, for $u_i \in G$, $i=1,2,\ldots,n$, and $VK_G = PK_n$. Then, he performs the following operations:

Step 1. Compute k_i by the following equation, for $i=1,2,\ldots,n$.

$$k_i = PK_i^{-1} \cdot r \bmod \omega . \quad (1)$$

Step 2. Select a random number k_0, where $k_0 \in Z_\omega^*$.
Step 3. Randomly select $sk1_i \in Z_\omega^*$, for $i = 1,2,...,n$.
Step 4. Calculate $sk2_i$ by the equation below, for $i = 1,2,...,n$.

$$sk2_i = k_i - k_{i-1} \cdot sk1_i \bmod \omega \quad (2)$$

Step 5. Securely distribute $sk_i = (sk1_i, sk2_i)$ to $u_i \in G$, for $i = 1,2,...,n$.
Step 6. Compute and deliver $w = \alpha^{k_0} \bmod N$ to u_1.

Multisignature generation phase:

Suppose $G = \{u_1, u_2, ..., u_n\}$, with signing structure $\mathrm{SER}[u_1, u_2, ..., u_n]$, want to generate a multisignature *MS* for a message *m*. Each u_i, for $i = 1,2,...,n$, performs the signing operations as below.

Step 1. Verify the partial multisignature S_{i-1} signed by u_{i-1}, (for $i \neq 1$) by testing if

$$(S_j)^{PK_j} = \beta^{h(m)} (\bmod N) . \quad (3)$$

(If the test fails, then the signing process is stopped and u_{i-1} is reported as a malicious signer.)

Step 2. Compute the partial multisignature S_i by

$$S_i = S_{i-1}^{sk1_i} \cdot \alpha^{h(m) \cdot sk2_i} \bmod N , \quad (4)$$

where S_{i-1} is generated by u_{i-1} and $S_0 = w^{h(m)} \bmod N$.

Step 3. Send S_i to u_{i+1}, for $i < n$.

The partial multisignature S_n generated by the last signer u_n is treated as the multisignature *MS* generated by *G* with $\mathrm{SER}[u_1, u_2, ..., u_n]$ for message *m*.

Multisignature verification phase:

The multisignature *MS* of message *m* that signed by the signing group *G* with signing structure $\mathrm{SER}[u_1, u_2, ..., u_n]$ can be publicly verified by VK_G as:

$$(MS)^{VK_G} = \beta^{h(m)} (\bmod N) . \quad (5)$$

LEMMA 1. For any message m and its partial multisignature S_i generated by $u_i \in G$, $S_i = \alpha^{k_i \cdot h(m)} (\bmod N)$ in the serial approach.
Proof:

Multiplying $h(m)$ to Equation 2 and raising both sides of it to exponents with base α, it yields a recursive relation $\alpha^{k_i \cdot h(m)} = \alpha^{k_{i-1} \cdot h(m) \cdot sk1_i} \cdot \alpha^{h(m) \cdot sk2_i} (\bmod N)$, where the $k_0 \in Z_\omega^*$ is randomly chosen. By the above fact and $\alpha^{k_0 \cdot h(m)} = w^{h(m)} = S_0 (\bmod N)$, we can conclude $S_i = \alpha^{k_i \cdot h(m)} (\bmod N)$ by mathematical inductions. Q.E.D.

THEOREM 1. If all $u_i \in G$ honestly sign the message m by following $\text{SER}[u_1, u_2, \ldots, u_n]$, then the generated multisignature MS will be successfully verified by Equation 5.
Proof:

Recall that $MS = S_n$ and $VK_G = PK_n$. By Equation 1 and Lemma 1 we can obtain Equation 5

$$(MS)^{VK_G} = \alpha^{k_n \cdot h(m) \cdot VK_G} = \alpha^{r \cdot PK_n^{-1} \cdot h(m) \cdot VK_G} = \beta^{h(m)} (\bmod N). \qquad \text{Q.E.D.}$$

THEOREM 2. Any disorder signing operation regarding $\text{SER}[u_1, u_2, \ldots, u_n]$ will be identified with the probability of $(\omega - 1)/\omega$.
Proof:

By following $\text{SER}[u_1, u_2, \ldots, u_n]$, u_i should sign the partial multisignature S_{i-1} generated by u_{i-1} for message m after verifying S_{i-1}'s validity. Assume a disorder operation takes place before u_i signs, whether by mistake or intentionally, that u_j, where $i < j \le n$, signs S_{i-1} instead of u_i. Then, the partial multisignature generated hereby will be

$$S_i' = S_{i-1}{}^{sk1_j} \cdot \alpha^{h(m) \cdot sk2_j} (\bmod N).$$

For S_i' to be successfully verified by Equation 5, it has to satisfy that $S_i' = S_i (\bmod N)$, which implies

$$S_{i-1}{}^{sk1_j} \cdot \alpha^{h(m) \cdot sk2_j} = S_{i-1}{}^{sk1_i} \cdot \alpha^{h(m) \cdot sk2_i} (\bmod N). \tag{6}$$

By Lemma 1, the exponent part of Equation 6 indicates

$$k_{i-1} \cdot sk1_j + sk2_j = k_{i-1} \cdot sk1_i + sk2_i (\bmod \omega). \tag{7}$$

In order for S_i' to be valid, two distinct private keys (i.e. sk_i and sk_j) have to satisfy Equation 7. Since the values of $sk1_i$'s for all $u_i \in G$ are randomly selected and $sk2_i$'s are computed from Equation 2, it is to see that the probability for sk_i and sk_j to satisfy Equation 7 is $1/\omega$. Therefore, the probability for successfully identifying a disorder event is $(\omega-1)/\omega$. Q.E.D.

3.2 For parallel signing structure

Let $G=\{u_1,u_2,...,u_n\}$ be a group of signers with signing structure $\mathrm{PAR}[u_1,u_2,...,u_n]$. The scheme is stated as follows.

Key Generation Phase:

SA prepares the partial verification keys and public verification key as $PK_i = f(ID_i)$, for $u_i \in G$, $i=1,2,...,n$, and $VK_G = f(\zeta(ID_1,ID_2,...,ID_n))$. Then, he performs the following operations:

Step 1. Compute k_i by the following equation, for $i=1,2,...,n$.

$$k_i = PK_i^{-1} \cdot r \bmod \omega. \tag{8}$$

Step 2. Select a random number k_0, where $k_0 \in Z_\omega^*$.

Step 3. Randomly select $sk1_i \in Z_\omega^*$ for $i=1,2,...,n$.

Step 4. Calculate the value of $sk2_i$ as follows, for $i=1,2,...,n$.

$$sk2_i = k_i - k_0 \cdot sk1_i \bmod \omega.$$

Step 5. Securely distribute $sk_i = (sk1_i, sk2_i)$ to $u_i \in G$, for $i=1,2,...,n$.

Step 6. Calculate the value of v by

$$v = \alpha^{VK_G^{-1} \cdot r - \sum_{i=1}^{n} k_i} \bmod N. \tag{9}$$

Step 7. Compute $w = \alpha^{k_0} \bmod N$ and deliver w, v to all $u_i \in G$.

Multisignature generation phase:

Suppose the signing group G with signing structure $\mathrm{PAR}[u_1,u_2,...,u_n]$ wants to generate the multisignature *MS* for message *m*. Each $u_i \in G$ performs the following tasks without concerning other signer's signing order.

Step 1. Compute the partial multisignature S_i as

$$S_i = (w^{h(m)})^{sk1_i} \cdot \alpha^{h(m) \cdot sk2_i} \bmod N.$$

Step 2. Send S_i to all $u_j \in G$, for $j \neq i$.
Step 3. Verify S_j sent from u_i, for $j \neq i$, by testing if

$$(S_j)^{PK_j} = \beta^{h(m)} (\bmod N). \tag{10}$$

(If the test fails, then the signing process is stopped and u_j is reported as a malicious signer.)
Step 4. Calculate the multisignature MS after receiving and verifying all S_j's, for $j \neq i$, $u_j \in G$, by the following equation.

$$MS = v^{h(m)} \cdot (\prod_{i=1}^{n} S_i) \bmod N. \tag{11}$$

Multisignature verification phase:

The multisignature *MS*, generated by the signing group *G* with signing structure PAR$[u_1, u_2, \ldots, u_n]$, for message *m* can be verified by testing if

$$(MS)^{VK_G} = \beta^{h(m)} (\bmod N). \tag{12}$$

THEOREM 3. If all $P_i \in G$, for $i = 1, 2, \ldots, n$, honestly sign the message *m* by following PAR$[u_1, u_2, \ldots, u_n]$, then the multisignature *MS* generated by *G* will be successfully verified by Equation 12.

Proof:

Based on the fact that all valid partial multisignatures can be successfully verified by Equation 10, we can rewrite Equation 10 with Equation 8 as

$$S_i = \alpha^{k_i \cdot h(m)} (\bmod N). \tag{13}$$

Then, from Equation 9, 11 and 13, we can obtain that

$$MS = v^{h(m)} \cdot (\prod_{i=1}^{n} S_i)(\bmod N)$$
$$= (\alpha^{VK_G^{-1} \cdot r - \sum_{i=1}^{n} k_i})^{h(m)} \cdot (\prod_{i=1}^{n} \alpha^{k_i})^{h(m)} (\bmod N)$$
$$= (\alpha^{VK_G^{-1} \cdot r - \sum_{i=1}^{n} k_i} \cdot \alpha^{\sum_{i=1}^{n} k_i})^{h(m)} (\bmod N)$$
$$= \alpha^{VK_G^{-1} \cdot r \cdot h(m)} (\bmod N)$$
$$= \beta^{VK_G^{-1} \cdot h(m)} (\bmod N).$$

This implies a valid *MS* will be successfully verified by Equation 12. Q.E.D.

3.3 For mixed signing structure

Assume $G = \{u_1, u_2, ..., u_n\}$ is a group consists of mixed-ordered signers. In any real case, the partial verification keys PK_i 's and the public verification key VK_G can be easily computed by following the rules described in section 2.2. The diagram representation of the signing structure is employed here to facilitate the key generation and multisignature generation phases.

A new notation used here is $prev(x)$, where x is a node in the signing structure diagram and $prev(x)$ indicates the set of nodes that directly connect and point to node x in the diagram.

Key generation phase:

By observing the diagram of the signing structure of *G*, *SA* first prepares the partial verification keys and public verification key, and then generates the secret key sk_i for each $u_i \in G$ as follows.

Step 1. Compute k_i by $k_i = PK_i^{-1} \cdot r \bmod \omega$, for $i = 1,2,...,n$.
Step 2. Select a random number k_0, where $k_0 \in Z_\omega^*$.
Step 3. Randomly select $sk1_i$, such that $sk1_i \in Z_\omega^*$, for $i = 1,2,...,n$.
Step 4. Calculate the value of $sk2_i$ for each $u_i \in G$ as follows.
If $prev(u_i) = \{s\}$, then $sk2_i = k_i - k_0 \cdot sk1_i \bmod \omega$;
Otherwise,

$$sk2_i = k_i - (\sum_{u_j \in prev(u_i)} k_j) \cdot sk1_i \bmod \omega.$$

Step 5. Distribute $sk_i = (sk1_i, sk2_i)$ to each $u_i \in G$ via a secure channel.

Step 6. Calculate a value v as follows.
If $|prev(t)|=1$, then $v=1$;
Otherwise,

$$v=\alpha^{VK_G^{-1}\cdot r-\sum_{u_j\in prev(t)}k_j} \bmod N.$$

Step 7. Compute and deliver $w=\alpha^{k_0} \bmod N$ to all u_i, for $prev(u_i)=\{s\}$.
Step 8. Send v to all $u_i\in G$.

Multisignature generation phase:

Supposing the signing group $G=\{u_1,u_2,\ldots,u_n\}$ with a mixed signing structure wants to generate a multisignature *MS* for a message *m*. Then, with the help of the corresponding diagram, each $u_i\in G$ performs the following operations to compute and distribute the partial multisignature.

Step 1. Compute the partial multisignature S_i as:
For u_i with $prev(P_i)=\{s\}$,

$$S_i=(w^{h(m)})^{sk1_i}\cdot\alpha^{h(m)\cdot sk2_i} \bmod N\text{; and,}$$

for u_i with $prev(u_i)\neq\{s\}$,

$$S_i=(\prod_{u_j\in prev(u_i)}S_j)^{sk1_i}\cdot\alpha^{h(m)\cdot sk2_i} \bmod N,$$

where S_j is the partial multisignature generated by $u_j\in prev(u_i)$.

Step 2. Distribute S_i to all $u_j\in G$, for $prev(u_j)\supseteq\{u_i\}$, and to all for $\{u_i,u_k\}\subseteq prev(t)$.

Afterwards, the multisignature *MS* for message *m* can be calculated by any $u_k\in prev(t)$ with the following equation:

$$MS=v^{h(m)}\cdot(\prod_{u_k\in prev(t)}S_k) \bmod N.$$

Note that before u_i signs, he should have verified the validity of each received partial multisignature S_j for $u_j\in prev(u_i)$ by testing if

$$(S_j)^{PK_j}=\beta^{h(m)} (\bmod N).$$

If the test fails, the signing process is stopped and the corresponding u_j is reported as a malicious signer.

Multisignature verification phase:

The multisignature *MS*, generated by the signing group *G* for the message *m* can be publicly verified as below.

$$(MS)^{VK_G} = \beta^{h(m)} (\bmod N).$$

4. SECURITY ANALYSIS

Possible attacks to the proposed schemes include the attempts to disclose the signer's private key and to forge a structured multisignature. Although the proposed schemes solve three different signing structures, they adopted the same techniques for key generation, multisignature generation and multisignature verification. Hereby we will show that the proposed schemes are secure against these attacks by focusing our discussion on the serial approach. Note that the security of the proposed schemes relies on the difficulty of solving discrete logarithm modulo a large composite (DLMC) [1] and factorising a large composite (FAC) [1, 17].

ATTACK 1. *An attacker attempts to reveal the secret key* $sk_i = (sk1_i, sk2_i)$ *of a signer* $u_i \in G$ *from all available public information.*
Analysis :

From Equation 1 and 2, it is to see that the secret key sk_i of u_i would be disclosed by the attacker only when he knows either the values of ω, *r*, and all PK_i^{-1}'s; or the values of ω and all k_i's. However, given all public information α, β, *N* and all $PK_{G,i}$'s for $u_i \in G$, computing ω from *N* is a problem of FAC intractability and deducing *r* from β is a difficulty of solving the problem of DLMC. In addition, the attacker may try to deduce the value of k_i from the result in Lemma 1, i.e. $S_i = \alpha^{k_i \cdot h(m)} (\bmod N)$. However, he will obviously face the problem of the DLMC intractability. □

ATTACK 2. *An attacker attempts to reveal the private key* $sk_i = (sk1_i, sk2_i)$ *of a signer* $u_i \in G$ *from the partial multisignatures* S_i*'s (for all* $u_i \in G$*) of a message m.*
Analysis:

Given $h(m)$, α , β , N and all S_i 's for $u_i \in G$, directly computing $(sk1_i, sk2_i)$ from Equation 4 in the serial ordered multisignature scheme is an intractability of the DLMC problem. On the other hand, solving $(sk1_i, sk2_i)$ from Equation 2 is also infeasible since ω and all k_i for $P_i \in G$ are secret parameters and known only to SA. □

ATTACK 3. *An attacker attempts to directly forge a valid multisignature for some message m for the signing group* $G = \{u_1, u_2, ..., u_n\}$.
Analysis
Since the private key of each $u_i \in G$ is securely kept, an attacker cannot create any partial multisignature or multisignature for some message *m* via Equation 4. Moreover, we know that a forged multisignature has to satisfy Equations 5 to be valid. However, with public information *N*, β , VK_G and $h(m)$, it's obviously that the attacker will face the FAC problem to directly solve *MS* from Equations 5. □

5. CONCLUSIONS

In this paper, we have addressed a new approach to multisignature schemes that applicable for various signing structures based on ID-based public keys. In addition to enforce the requirement that all signers in the signing group have to follow the predefined signing structure when generating a multisignature, our scheme has the merits that both the length of multisignature and the computation effort for multisignature verification are fixed and independent to the amount of signers. Due to the intractability of the DLMC problem and the FAC problem, the proposed scheme is secure against the deduction of the signer's secret key and forgery to the multisignature.

ACKNOWLEDGEMENT

The authors would like to thank the anonymous referees for their useful comments on improving our paper. This work was partially supported by the Ministry of Education, Taiwan, Program of Excellence Research 90-E-FA04-1-1.

REFERENCES

[1] L.M. Adleman and K.S. McCurley, "Open problems in number-theoretic complexity, II", *Proc. First Algorithmic Number Theory Symposium*, Springer-Verlag, 1994, pp.291-322.

[2] C. Boyd, "Digital Multisignatures", *IMA Conference on Cryptography and Coding*, Oxford University Press, 1989, pp. 241-246.

[3] C. Boyd, "Multisignatures based on Zero Knowledge Schemes", *Electronics Letters*, 27 (22), October 1991, pp. 2002-2004.

[4] M. Burmester, Y. Desmedt, H. Doi, M. Mambo, E. Okamoto, M. Tada and Y. Yoshifuji, "A structured ElGamal-type multisignature scheme", *Proc. Workshop on Practice and Theory in Public Key Cryptosystems*, LCNS 1751, Springer-Verlag, 2000, pp. 466-483.

[5] Y.S. Chang, T.C. Wu and S.C. Huang, "ElGamal-like digital signature and multisignature schemes using self-certified public keys", *The Journal of Systems and Software*, 50 (2), 2000, pp. 99-105.

[6] H. Doi, E. Okamoto and M. Mambo, "Multisignature schemes for various group structures", *The 36-th Annual Allerton Conference on Communication, Control, and Computing*, 1999, pp. 713-722.

[7] A. Fiat and A. Shamir, "How to prove yourself: practical solution to identification and signature problems", *Advances in Cryptology – CRYPTO'86*, Springer-Verlag, 1987, pp. 186-194.

[8] T. Hardjono and Y. Zheng, "A practical digital multisignature scheme based on discrete logarithms", *Advance in Cryptology–AUSCRYPT'92*, Springer-Verlag, 1993, pp.122-132.

[9] L. Harn and T. Kielser, "New scheme for digital multisignatures", *Electronics Letters*, 25 (15), 1989, pp. 1002-1003.

[10] K. Itakura and K. Nakamura, "A public-key cryptosystem suitable for digital multisignature", *NEC Research and Development*, Vol. 71, October 1983, pp. 1-8.

[11] M. Mambo, K. Usuda and E. Okamoto, "Proxy signatures: Delegation of the power to sign messages", *IEICE Tran. Fundamentals*, E97-A (9), 1996, pp. 1338-1353.

[12] M. Mambo, K. Usuda and E. Okamoto, "Proxy signatures for delegating signing operation", *Proc. Conf on Computer and Comm. Security*, ACM press 1996, pp. 48-57.

[13] K. Ohta, S. Micali and L. Reyzin, "Accountable-subgroup Multisignatures", *Manuscript*, Massachusetts Institute of Technology, Cambridge, MA, USA, Nov. 2000.

[14] K. Ohta and T. Okamoto, "A digital multisignature scheme based on the Fiat-Shamir scheme", *Advance in Cryptology-ASIACRYPT'91*, Springer-Verlag, 1993, pp. 139-148.

[15] K. Ohta and T. Okamoto, "Multisignature schemes secure against active insider attacks", *IEICE Transactions on Fundamentals*, E82-A (1), 1999, pp. 21-31.

[16] T. Okamoto, "A digital multisignature scheme using bijective public-key cryptosystems", *ACM Tran. Computer Systems*, 6 (8), 1988, pp. 432-441.

[17] R.L. Rivest, A. Shamir and L. Adleman, "A method for obtaining digital signatures and public key cryptosystems", *Comm. of the ACM*, 21 (2), 1978, pp. 120-126.

[18] A. Shamir, "Identity-based cryptosystems and signature schemes", *Advances in Cryptology – CRYPTO'84*, Springer-Verlag, 1985, pp. 47-53.

[19] T.C. Wu, S.L. Chou and T.S. Wu, "Two ID-based multisignature protocols for sequential and broadcasting architectures", *Computer Comm.* 19 (9-10), 1996, pp. 851-856.

PROBABILISTIC RELATIONS FOR THE SOLITAIRE KEYSTREAM GENERATOR

Marina Pudovkina
Moscow Engineering Physics Institute
maripa@online.ru

Abstract: Stream ciphers are often used in applications where high speed and low delay are a requirement. The Solitaire keystream generator was developed by B. Schneier as a paper-and-pencil cipher. Solitaire gets its security from the inherent randomness in a shuffled deck of cards. In this paper we present probabilistic relations for the Solitaire keystream generator and describe their properties.

Keywords: Solitaire. Probabilistic relations.

1. INTRODUCTION

Many keystream generators proposed in the literature consist of a number of possibly clocked linear feedback shift registers (LFSEs) that are combined by a function with or without memory. LFSR-based generators are often hardware oriented and for a variety of them it is known how to achieve desired cryptographical properties [6]. For software implementation, a few keystream generators have been designed which are not based on shift registers. Such generators with mixing next-state functions are RC4 [7], IA, IBAA, ISAAC [8], SCOP [9].

The Solitaire keystream generator was developed by B. Schneier [1] as a paper-and-pencil cipher. Solitaire gets its security from the inherent randomness in a shuffled deck of cards. By manipulating this deck, a communicant can create a string of "random" letters that he then combines with his message. Solitaire can be simulated on a computer, but it is designed to be implemented by hand. It was designed to be secure even against the most well–funded military adversaries with the biggest computers and the smartest cryptanalysts. It's not fast, though it can take an evening to encrypt or decrypt a reasonably long message.

Solitaire is an output–feedback mode stream cipher. The next–state function F is a composition of four transformations $F = F_4 F_3 F_2 F_1$ which permute of elements of a deck.

In [2] is considered cycle structure of Solitaire. It is proved that Solitaire is not reversible and described all irreversible states. In [3] are analyzed properties of the key scheduling algorithm which derives the initial state from a variable size key, and described weaknesses of this process. One of these weaknesses is the existence of large classes of equivalent keys.

In this paper we present probabilistic relations for the Solitaire keystream generator and stress some their properties. These relations describe the jokers location in a deck at any time t. We show that the number of elements between the jokers at time t depends on t and the initial number of elements between the A joker and the B joker.

The paper is organized in the following way. In section 2 we describe the Solitaire cipher. In section 3 we consider probabilistic relations for the next-state function and in section 4 we give them for the key scheduling algorithm. We conclude in section 5.

2. DESCRIPTION OF THE SOLITAIRE CIPHER

Solitaire is in fact a family of algorithms indexed by parameter n, which is a positive integer. Let m be a cardinality of an alphabet of a plaintext, then n=2m+2. The internal state of Solitaire at time t consists of a table $S_t=\{s_t[0],\ldots,s_t[n-1]\}$ of n values. S is a permutation of integers between zero and n–1.

B. Schneier takes n=54, m=26.

The next–state function F is a composition of four transformations $F = F_4 F_3 F_2 F_1$, which correspond to items 1–4 of the description given in [1]. The transformations F_4, F_3, F_2, F_1 permute elements of a table $S=(s[0],\ldots,s[n-1])$.

Let one joker A=n–2 and the other B=n–1.

The next–state function F

1. The transformation F_1: $S_i \rightarrow X = (x[0],\ldots,x[n-1])$. Let $s[j]=n-1(A)$. If $j \neq n-1$ then move the A joker one element down: $x[j]=s[j+1]$, $x[j+1]=n-1$, and $x[k]=s[k]$, $k=0\ldots n-1$, $k \neq j, j+1$. If $j=n-1$ move it just below s[0]: $x[0]=s[0]$, $x[1]=A$, $x[2]=s[1],\ldots, x[k]=s[k-1],\ldots,x[n-1]=s[n-2]$.

2. The transformation F_2: X→Y=(y[0],...,y[n–1]). Let x[j]=n–2(B). If j≠ n–1 and n–2 then move the B joker two elements down: y[j]=x[j+1], y[j+1]=x[j+2], y[j+2]=n–2 (B). If j=n–1, move the B joker just below x[1]. If j=n–2, move it just below x[0].
3. The transformation F_3: Y → Z=(z[0],...,z[n–1]). Perform a triple cut. That is, swap the elements above the first joker with the elements below the second joker. "First" and "second" jokers refer to whatever joker is nearest to, and furthest from, the top of the deck. Ignore the "A" and "B" designations for this step. The jokers and the elements between them don't move; the other elements move around them.
4. The transformation F_4: Z → S_{i+1}. Perform a count cut. Let z[n–1]=k. Swap the elements z[0],...,z[k] with the elements z[k+1],..., z[n–2]. The element z[n–1] does not swap. A deck with a joker as z[n–1] will remain unchanged by this step.

The output function f

Let $s_{i+1}[0]=q$.

If $s_{i+1}[0]=A$ or $s_{i+1}[0]=B$ then we have not an output element.

If $s_{i+1}[0]\neq A$, B then the output element $k_i=s_{i+1}[q]$ (mod m).

Let $M=m_1m_2\ldots m_L$ be a plaintext and $C=c_1c_2\ldots c_L$ be a ciphertext.

Encryption:

$$c_i=(m_i+k_i) \pmod m.$$

Decryption:

$$k_i=(c_i-m_i) \pmod m.$$

Key Scheduling Algorithm

Key is an initial deck ordering. A passphrase is used to order the deck. This method uses the Solitaire algorithm to create an initial deck ordering. Both the sender and receiver share a passphrase. (For example, "SECRET KEY.") Start with the deck in a fixed order; (0, 1, 2, ...,n–3, A, B). Perform the Solitaire operation, but instead of Step 4, do another count cut based on the first character of the passphrase. In other words, do step 4 a second time, using the character of the passphrase as the cut number instead of the last card.

Repeat the four steps of the Solitaire algorithm once for each character of the passphrase. That is, the second time through the Solitaire steps use the second character of the passphrase, the third time through use the third character, etc. Use the final two characters to set the positions of the jokers.

3. PROBABILISTIC RELATIONS FOR THE NEXT-STATE FUNCTION

In this section we describe probabilistic relations and their properties for the next-state function. Let d_A be the number of elements before the A joker and d_B be the number of elements before the B joker. We shall say that d_A is called the A joker distance and d_B is called the B joker distance.

THEOREM 1

Let $d_A(j)$ be the A joker distance at time j and $d_B(j)$ be the B joker distance at time j. Let $S_j=(\boldsymbol{F})^j(S_0)$ and $k_j=s_j[n-1]$. If for any t<j: $d_B[t]\neq n-1$, $d_B[t]\neq n-2$, $d_A[t]\neq n-1$, $d_B[t]\neq d_A[t]+1$ and $d_A[t]\neq d_B[t]+1$, then the A joker distance and the B joker distance satisfy the following relations.

1. If j=2i then

$$d_A(2i)=[d_A(0)+\sum_{j=0}^{i-1}(k_{2j+1}-k_{2j+2})-i]\ (\mathrm{mod}\ n-1)$$

$$d_B(2i)=[d_B(0)+\sum_{j=0}^{i-1}(k_{2j+1}-k_{2j+2})+i]\ (\mathrm{mod}\ n-1) \tag{1}$$

2. If j=2i+1 then

$$d_A(2i+1)=[-d_B(0)+\sum_{j=0}^{i-1}(k_{2j+2}-k_{2j+1})-(1+k_{2i+1})(\mathrm{mod}\ n)-i-2]\ (\mathrm{mod}\ n-1)$$

$$d_B(2i+1)=[-d_A(0)+\sum_{j=0}^{i-1}(k_{2j+2}-k_{2j+1})+i-1-(1+k_{2i+1})(\mathrm{mod}\ n)]\ (\mathrm{mod}\ n-1). \tag{2}$$

Proof.

We conduct the proof by induction. Let d'_A be the A joker distance in a permutation Y and d'_B be the B joker distance in a permutation Y. Let d''_A be the A joker distance in a permutation Z and d''_B be the B joker distance in a permutation Z. Consider j=2k+1.

Let us remark that

$$d'_A(2k+1)=d_A(2k)+1,$$

$$d'_B(2k+1)=d_B(2k)+2.$$

Recall that $F_2\ F_1(S_{2k})$ =Y and $F_3(Y)$ =Z. Note that the distances in permutations S_{2k}, Y, and Z satisfy the following relations.

$$d_A''(2k+1)=n-1-d'_B(2k+1)=n-3-d_B(2k),$$

$$d_B''(2k+1)=n-1-d'_A(2k+1)=n-2-d_A(2k).$$

The distances in permutations S_{2k},Y, Z and S_{2k+1} satisfy the following relations.

$d_A(2k+1)=[d_A''(2k+1)-(k_{2k+1}+1) \pmod n] \pmod{n-1}=[n-1-d'_B(2k+1)-(k_{2k+1}+1) \pmod n] \pmod{n-1}= [-d'_B(2k+1)-(k_{2k+1}+1)] \pmod{n-1}= [-d_B(2k)-2-(k_{2k+1}+1)] \pmod{n-1}$,

$d_B(2k+1)=[d_B''(2k+1)-(k_{2k+1}+1) \pmod n] \pmod{n-1}= [n-1-d'_A(2k+1)-(k_{2k+1}+1) \pmod n] \pmod{n-1}=[-d'_A(2k+1)-(k_{2k+1}+1) \pmod n] \pmod{n-1} = [-d_A(2k)-1-(k_{2k+1}+1) \pmod n] \pmod{n-1}$.

It follows that

$$d_A(2k+1)= -[d_B(2k)+(1+k_{2k+1}) \pmod n+2] \pmod{n-1},$$

$$d_B(2k+1)= -[d_A(2k)+(1+k_{2k+1}) \pmod n+1] \pmod{n-1}.$$

Therefore,

$d_A(2k+2)=-[d_B(2k+1)+(1+k_{2k+2}) \pmod n+2] \pmod{n-1}= -[-d_A(2k)-(1+k_{2k+1}) \pmod n-1+(1+k_{2k+2}) \pmod n+2] \pmod{n-1}= [-d_A(2k)+(k_{2k+2}-k_{2k+1})+1] \pmod{n-1}= [d_A(2k)+(k_{2k+1}-k_{2k+2})-1] \pmod{n-1}$,

$d_B(2k+2)=-[d_A(2k+1)+(1+k_{2k+2}) \pmod n+1] \pmod{n-1}= -[-d_B(2k)-(1+k_{2k+1}) \pmod n-2+(1+k_{2k+2}) \pmod n+1] \pmod{n-1}=-[-d_B(2k)+(k_{2k+2}-k_{2k+1})-1] \pmod{n-1}= [d_B(2k)+(k_{2k+1}-k_{2k+2})+1] \pmod{n-1}$.

We apply an induction over k and obtain.

$d_A(2k+2)=[d_A(2k) + (k_{2k+1}-k_{2k+2})-1] \pmod{n-1}= [d_A(2k-2)+(k_{2k-1}+k_{2k+1}-k_{2k}-k_{2k+2})-2] \pmod{n-1}=\ldots=[d_A(0)+\sum_{j=0}^{k} (k_{2j+1} - k_{2j+2})-k+1] \pmod{n-1}$.

$d_B(2k+2)= [d_B(2k) + (k_{2k+1}-k_{2k+2})+1] \pmod{n-1}=[d_B(2k-2)+(k_{2k-1}+k_{2k+1}-k_{2k}-k_{2k+2})+2] \pmod{n-1}=\ldots=[d_B(0)+ \sum_{j=0}^{k} (k_{2j+1} - k_{2j+2})+k+1] \pmod{n-1}$.

$d_A(2k+1)=-[d_B(2k)+(1+k_{2k+1}) \pmod{n}+2] \pmod{n-1}= -[d_B(0)+ \sum_{j=0}^{k-1} (k_{2j+1}-k_{2j+2})+k+2+(1+ k_{2k+1}) \pmod{n}] \pmod{n-1}= -[d_B(0)+ \sum_{j=0}^{k-1} (k_{2j+2}-k_{2j+1})-k-2-(1+k_{2k+1}) \pmod{n}] \pmod{n-1}$.

$d_B(2k+1)=-[d_A(2k)+(1+k_{2k+1}) \pmod{n}+1] \pmod{n-1}= -[d_B(0)+ \sum_{j=0}^{k-1} (k_{2j+1}-k_{2j+2})-k+1+(1+k_{2k+1}) \pmod{n}] \pmod{n-1}=[-d_B(0)+ \sum_{j=0}^{k-1} (k_{2j+2}-k_{2j+1})+k-1-(1+k_{2k+1}) \pmod{n}] \pmod{n-1}$.

This completes the proof.

REMARK 1

Let $P\{d_B=n-1, d_B=n-2, d_A=n-1, d_B= d_A +1, d_A= d_B +1\}$ be a probability that $d_B=n-1$, or $d_B=n-2$, or $d_A=n-1$, or $d_B= d_A +1$, or $d_A= d_B +1$ then

$$P\{d_B=n-1,d_B=n-2,d_A=n-1,d_B= d_A +1,d_A= d_B +1\}\leq 4/n.$$

Proof.

Really, $P\{d_B=n-1,d_B=n-2,d_A=n-1,d_B= d_A +1,d_A=d_B +1\}\leq 3\cdot(n-1)!/n! +2\cdot(n-2)!/n!=3/n+2/(n-1)n\leq 4/n$.

Let Prob(j) be a probability that the probabilistic relations at time j are true.

REMARK 2

$Prob(j) \geq (1-4/n)^j$.

Proof.

Note that the probabilistic relations at time j are true if for any t<j: $d_B[t] \neq n-1$, $d_B[t] \neq n-2$, $d_A[t] \neq n-1$, $d_B[t] \neq d_A[t]+1$ and $d_A[t] \neq d_B[t]+1$. Using remark 1, we have $P \geq (1-4/n)^j$.

Let us consider some properties that obtained from the presented probabilistic relations. By $dist_{AB}(t)$ denote the number of elements between the A joker and the B joker at time t. Proposition 1 and proposition 2 show that $dist_{AB}(t)$ depends on t and $d_A(0) - d_B(0)$.

PROPOSITION 1

Let $x = (d_A(i) - d_B(i)) \pmod{n-1}$ then

$$dist_{AB}(i) \in \{x-1, n-2-x\}$$

Proof.

Let us remark that $dist_{AB}(i) = |d_A(i) - d_B(i)| - 1$.

Consider two possible cases.

a) If $d_A(i) > d_B(i)$ then $dist_{AB}(i) = (d_A(i) - d_B(i)) \pmod{n-1} - 1 = x-1$

b) If $d_A(i) < d_B(i)$ then $x = (d_A(i) - d_B(i)) \pmod{n-1} = n-1 + d_A(i) - d_B(i)$. Therefore, $dist_{AB}(i) = d_B(i) - d_A(i) - 1 = n-2-X$

The proposition is proved.

PROPOSITION 2

Let $y = (d_A(0) - d_B(0) - k) \pmod{n-1}$ then

$$dist_{AB}(k) \in \{y-1, n-2-y\}$$

Proof.

Let $x = (d_A(i) - d_B(i)) \pmod{n-1}$. By proposition 1 and (1), (2) we obtain.

a) If k=2i+1 then

$$x = (d_A(k) - d_B(k)) \pmod{n-1} = (-d_B(0) + \sum_{j=0}^{i-1} (k_{2j+2} - k_{2j+1}) - (1 + k_{2i+1}) \pmod{n}$$

$$-i-2-(-d_A(0) + \sum_{j=0}^{i-1} (k_{2j+2} - k_{2j+1}) + i - 1 - (1 + k_{2i+1}) \pmod{n}))) \pmod{n-1} = \ldots$$

$$= (d_A(0) - d_B(0) - 2i - 1) \pmod{n-1} = (d_A(0) - d_B(0) - k) \pmod{n-1}.$$

b) If k=2i then

$$x = (d_A(k) - d_B(k)) (\mathrm{mod}\ n-1) = (d_A(0) + \sum_{j=0}^{i-1} (k_{2j+1} - k_{2j+2}) - i - d_B(0) + \sum_{j=0}^{i-1} (k_{2j+1} - k_{2j+2}) + i) (\mathrm{mod}\ n-1) = (d_A(0) - d_B(0) - 2i)(\mathrm{mod}\ n-1) = (d_A(0) - d_B(0) - k) (\mathrm{mod}\ n-1).$$

Therefore, $x=(d_A(k)-d_B(k))(\mathrm{mod}\ n-1)=(d_A(0)-d_B(0)-k)(\mathrm{mod}\ n-1) = y$.
We have $dist_{AB}(k) \in \{y-1, n-2-y\}$
The proposition is proved.

REMARK 3

If we take $x=(d_B(k)-d_A(k))(\mathrm{mod}\ n-1)$ and $y=(d_B(0)-d_A(0)+k)$ $(\mathrm{mod}\ n-1)$ then proposition 1 and proposition 2 remain true.

PROPOSITION 3

If $d_B(0)-d_A(0)=2$ then

$$dist_{AB}(k) \in \{k+1, n-k-4\}.$$

This proposition can be proved by direct calculations.

In propositions 4–7 we describe some properties which allow finding the jokers location or elements of the permutation with high probabilities. Let S'_0, S''_0 be two initial states. By d'_A, d'_B denote the A joker distance and the B joker distance for S' and by d''_A, d''_B denote distances for S''.

PROPOSITION 4

1. If $d'_A(0)= d''_A(0)$, $d'_B(0)= d''_B(0)$ and $d'_A(k)= d''_A(k)$ then $d'_B(k)= d''_B(k)$.
2. If $d'_A(0)= d''_A(0)$, $d'_B(0)= d''_B(0)$ and $d'_B(k)= d''_B(k)$ then $d'_A(k)=d''_A(k)$.

   Proof.

   Let us prove item 1.

a) Let $k=2i$.

   By $d'_A(k)= d''_A(k)$ and (1) we get

$$[d'_A(0) + \sum_{j=0}^{i-1} (k'_{2j+1} - k'_{2j+2}) - i] = [d''_A(0) + \sum_{j=0}^{i-1} (k''_{2j+1} - k''_{2j+2}) - i]\ (\mathrm{mod}\ n-1).$$

Therefore,

$$\sum_{j=0}^{i-1} (k'_{2j+1} - k'_{2j+2}) = \sum_{j=0}^{i-1} (k''_{2j+1} - k''_{2j+2})\ (\mathrm{mod}\ n-1).$$

Note that

$d'_B(2i)= [d'_B(0)+ \sum_{j=0}^{i-1} (k'_{2j+1}- k'_{2j+2})+i] \pmod{n-1}$ and $d''_B(2i)= [d''_B(0)+ \sum_{j=0}^{i-1} (k''_{2j+1} - k''_{2j+2})+i] \pmod{n-1}$.

Therefore, $d'_B(k)= d''_B(k)$.

b) Let $k=2i+1$.

From (2) and $d'_A(k)= d''_A(k)$ it follows that

$$[-d'_B(0)+\sum_{j=0}^{i-1} (k'_{2j+2} - k'_{2j+1})-(1+k'_{2i+1}) \pmod{n} - i-2] =$$

$$[-d''_A(0)+\sum_{j=0}^{i-1} (k''_{2j+2}- k''_{2j+1})+i-1-(1+k''_{2i+1}) \pmod{n}] \pmod{n-1}.$$

By

$$d'_B(k)= [-d'_A(0)+\sum_{j=0}^{i-1} (k'_{2j+2}- k'_{2j+1})+i-1-(1+k'_{2i+1}) \pmod{n}] \pmod{n-1},$$

$$d''_B(k)= [-d''_A(0)+ \sum_{j=0}^{i-1} (k''_{2j+2}-k''_{2j+1})+i-1-(1+k''_{2i+1}) \pmod{n}] \pmod{n-1}$$

we have

$\sum_{j=0}^{i-1} (k'_{2j+2}-k'_{2j+1})-(1+k'_{2i+1}) \pmod{n}= \sum_{j=0}^{i-1} (k''_{2j+2} - k''_{2j+1})+i-1-(1+k''_{2i+1}) \pmod{n}$. Therefore, $d'_A(k)= d''_A(k)$.

Item 2 is proved similarly.

The proposition is proved.

PROPOSITION 5

If we know either $\{d_A(k), d_B(k+1)\}$ or $\{d_B(k), d_A(k+1)\}$ then we can determine the value of k_{k+1}.

This proposition can be proved by direct calculations.

PROPOSITION 6

1. If we know either $\{d_A(0), d_A(2i)\}$ or $\{d_B(0), d_B(2i)\}$ then we can determine

$$I=\sum_{j=0}^{i-1}(k_{2j+1}-k_{2j+2}).$$

2. If we know either $\{d_B(0), d_A(2i+1)\}$or $\{d_A(0), d_B(2i+1)\}$ then we can determine

$$I=\sum_{j=0}^{i-1}(k_{2j+2}-k_{2j+1})-(1+k_{2i+1}).$$

Proof.
Let us prove item 1.
Note that (1) we can rewrite as

$$d_A(2i)=\sum_{j=0}^{i-1}(k_{2j+1}-k_{2j+2})+d_A(0)-i \pmod{n-1}=I+d_A(0)-i \pmod{n-1},$$

$$d_B(2i)=\sum_{j=0}^{i-1}(k_{2j+1}-k_{2j+2})+d_B(0)+i \pmod{n-1}=I+d_B(0)+i \pmod{n-1}.$$

This yields that

$$I=\sum_{j=0}^{i-1}(k_{2j+1}-k_{2j+2})=d_A(2i)-d_A(0)+i \pmod{n-1},$$

$$I=\sum_{j=0}^{i-1}(k_{2j+1}-k_{2j+2})=d_B(2i)-d_B(0)-i \pmod{n-1}$$

Item 2 is proved similarly.
The proposition is proved.

PROPOSITION 7

1. If $d_A(0)=d_A(k)$ then $d_B(k)=(d_B(0)+k) \pmod{n-1}$.
2. If $d_B(0)=d_B(k)$ then $d_A(k)=(d_A(0)-k) \pmod{n-1}$.

This proposition can be proved by direct calculations.

Let c be an element of the permutation S and $c \notin \{A, B\}$ and by $d_c(j)$ denote the number of elements before c at time j. In proposition 8 we find $d_c(1)$.

PROPOSITION 8

1. If either $d_A(0)< d_c(0)< d_B(0)$ or $d_B(0)< d_c(0)< d_A(0)$ then $d_c(1)= [-d_A(0) - d_B(0)-k_1+ d_c(0)-4] \pmod{n-1}$.
2. If either $d_c(0)< d_A(0)<d_B(0)$ or $d_B(0)< d_A(0)< d_c(0)$ then $d_c(1)= [d_c(0)-d_A(0)-k_1-2] \pmod{n-1}$.
3. If either $d_c(0)<d_B(0)<d_A(0)$ or $d_A(0)<d_B(0)< d_c(0)$ then $d_c(1)= [d_c(0)-d_B(0)-k_1-2] \pmod{n-1}$.

The proof is straightforward.

4. PROBABILISTIC RELATIONS FOR THE KEY SCHEDULING ALGORITHM

In this section we present probabilistic relations for the key scheduling algorithm.

THEOREM 2

Let $K=k_1,\ldots, k_L$ be a passphrase, where L is its length. Let $S_0=(0, 1, 2, \ldots, n-3, A, B)$ and $S_j=(F)^j(S_0)$. Let $d_A(j)$ be the A joker distance at time j and $d_B(j)$ be the B joker distance at time j. If for any $t<j$: $d_B[t]\neq n-1$, $d_B[t]\neq n-2$, $d_A[t] \neq n-1$, $d_B[t] \neq d_A[t]+1$ and $d_A[t] \neq d_B[t]+1$, then the A joker distance and the B joker distance satisfy the following relations.

1. If j=2i then

$$d_A(2i)=[d_A(0)+\sum_{j=0}^{i-1} (k_{2j+1} - k_{2j+2}) - i] \pmod{n},$$

$$d_B(2i)=[d_B(0)+\sum_{j=0}^{i-1} (k_{2j+1}- k_{2j+2})+i] \pmod{n}.$$

2. If j=2i+1 then

$$d_A(2i+1)= [-d_B(0)+\sum_{j=0}^{i-1} (k_{2j+2} - k_{2j+1})-k_{2i+1}-i-3] \pmod n,$$

$$d_B(2i+1)= [-d_A(0)+\sum_{j=0}^{i-1} (k_{2j+2} - k_{2j+1})+i-2-k_{2i+1}] \pmod n.$$

This theorem is proved as theorem 1.

We stress that the propositions which are proved in the previous section remain true for the key scheduling algorithm but the operation "mod(n-1)" is changed by "(mod n)".

5. CONCLUSION

In this paper we presented probabilistic relations for the jokers in Solitaire keystream generator and described some their properties. It analyzes the probability distribution of distance between the two jokers in a deck at different time periods. We found that the number of elements between jokers at time t depends on t and the initial number of elements between the A joker and the B joker. Presented results with bit changes are applied to the key-scheduling algorithm of Solitaire.

We hope that results described in [2], [3] and this paper allow mounting an attack on this cipher which is more effective than exhaustive search, and this will be the object of another paper.

REFERENCES

[1] Schneier B., "The Solitaire Encryption Algorithm", http://www.counterpane.com/solitaire.html.

[2] Pudovkina M, Varfolomeev A.A. "A Cycle Structure of the Solitaire Keystream Generator". 3nd International Workshop on Computer Science and Information Technologies CSIT'2001, YFA, 2001.

[3] Pudovkina M., "Weakness in the Key Scheduling Algorithm of the Solitaire Keystream Generator", SIBCOM-2001, TOMSK. (the paper is being published)

[4] Varfolomeev A.A., Zhukov A.E., Pudovkina M., "Analysis of Stream Ciphers ', Moscow, MEPhI, 2000.

[5] Crowley P. "Problems with Bruce Schneier "Solitaire""., http://www.hedonism.demon.co.uk/ paul/solitaire/

[6] Rueppel R.A. "Analysis and Design of Stream Ciphers", Springer-Verlag, Communications and Control Engineering Series, 1986.

[7] Rivest R.L.,"The RC4 encryption algorithm", RSA Data Security, Inc., Mar. 1992
[8] R.J. Jenkins, "ISAAC", Fast Software Encryption –Cambridge 1996, vol. 1039, D. Gollmann ed., Springer-Verlag.
[9] Simeon V. Maltchev and Peter T. Antonov, "The SCOP Stream Cipher", ftp://ftp.funet.fi/pub/crypt/cryptography/symmetric/scop/scop.tar.gz, Dec. 1997.

HAZARD ANALYSIS FOR SECURITY PROTOCOL REQUIREMENTS

Nathalie Foster*
Jeremy Jacob
Department of Computer Science
University of York
UK
{nathalie.foster, jeremy.jacob}@cs.york.ac.uk

Abstract This paper describes a process for the generation and analysis of security protocol requirements. It addresses some of the problems resulting from the inadequacies of present development methods. It is based on a hazard analysis technique which has been developed for safety critical systems engineering. This provides a structured method of analysis of the requirements whilst avoiding the problems of being too restrictive.

Keywords: security protocols, software engineering, requirements gathering and analysis, hazard analysis.

1. Introduction

1.1. Security Protocol Development

In comparison to the process of general software development, security protocol development is relatively unstructured. It is therefore hardly surprising that protocols are still being published that are later found to be vulnerable to attacks. Sound engineering practices need to be applied to protocol development and we need to consider the whole development lifecycle for protocols.

General models of software development have been proposed, the most famous of these are the Waterfall model by Royce[12] and the Spiral model by Boehm[5]. Many variations on these models have been suggested but common to all the models are a number of distinct activities: of these requirements gathering and analysis are the first [9, 13]. Al-

*Supported by an Engineering and Physical Sciences Research Council (EPSRC) studentship

though not perfect, these models go some way towards the development of higher quality software.

Research in protocol development has focused on the use of formal methods and logics in the design and verification of protocols at a late stage in the development. Little work has been carried out into the relevance to protocol development of other activities found in the software engineering process, such as requirements engineering.

We believe that many attacks and flaws could be guarded against through the proper gathering and analysis of the protocol requirements before the actual design. This would reduce the incidence of security failures due to the 'opportunistic exploitation of elementary design flaws' or 'implementation and management design errors' [1, 2].

In this paper, we address the issue of requirements gathering and analysis in protocol development. Our approach is based on a hazard analysis technique and is described in Section 1.2 and 2. It is used to analyse the goals of the protocol with reference to the security requirements of the protocol, in order to generate the low level protocol requirements.

Through the use of this technique, we are able to begin designing a protocol with a more thorough understanding about what it should do. We also have a higher level of confidence about the security of the protocol designed based on the requirements, before we carry out any verification of the protocol. This is necessary for the verification of the protocols: we have requirements which we can verify the protocol against, rather than having to guess what the requirements are before we can start verification.

1.2. Hazard Analysis

HAZOP. We propose the use of a hazard analysis technique which has its foundations in a method called Hazard and Operability Study (HAZOP) [6, 7]. HAZOP was developed by Imperial Chemical Industries (ICI) for the identification of hazards in process plant designs within the chemical industries, where the analysis is carried out on the pipework and instrumentation design of the plant. It has since been applied in the food-processing, pharmaceutical, nuclear, oil and gas industries and has also been adapted for use in the development of safety critical systems [11].

In a HAZOP study, a team identifies the entities and attributes of a design. A standard list of guide words is used to suggest deviations to these attributes. The deviations are analysed to determine their possible causes and effects and to consider what actions need to be taken to avoid or minimise the effects of the deviations. The results of the analysis are

GUIDE WORD	GENERIC MEANING
Omission	The service is never delivered, i.e. there is no communication. These are classified as either *total* or *partial.*
Commission	A service is delivered when not required, i.e. there is an unexpected communication. These are classified as either *spurious* or *repetition.*
Early	The service (communication) occurs earlier than intended. This may be *absolute* (i.e. early compared to a real-time deadline) or *relative* (early with respect to other events or communications in the system).
Late	The service (communication) occurs later than intended. As with early, this may be absolute or relative.
Value	The information (data) delivered has the wrong value.

Table 1. The SHARD guide words [10]

recorded in a table detailing the deviations, causes, effects, detection and protection, and the justification and recommendations. The analysis documentation is used to improve the safety of the system under study and may also be used in further investigation of the safety of the system.

SHARD. Software Hazard Analysis and Resolution in Design (SHARD) [10], is a 'projective computer system safety analysis technique based on HAZOP'. It is used to analyse designs and to obtain system safety related requirements for the detailed development of those designs. The guide words in SHARD are based on the communication of pieces of information, with specific values, at particular points in time (Table 1).

The analysis process in SHARD is even more structured than in HAZOP, with extra steps to be carried out in the analysis. The SHARD process is shown in the flow diagram in Figure 1.

The analysis is recorded in a table with at least the following column headings: *Guide word; Deviation; Possible Causes; Effects; Detection and Protection; Justification/Design Recommendations.*

The structured SHARD process and the more appropriate guide words for a system involving information flows lends itself to the analysis of security protocol requirements with some modifications. The application of SHARD to protocol requirements gathering and analysis is described

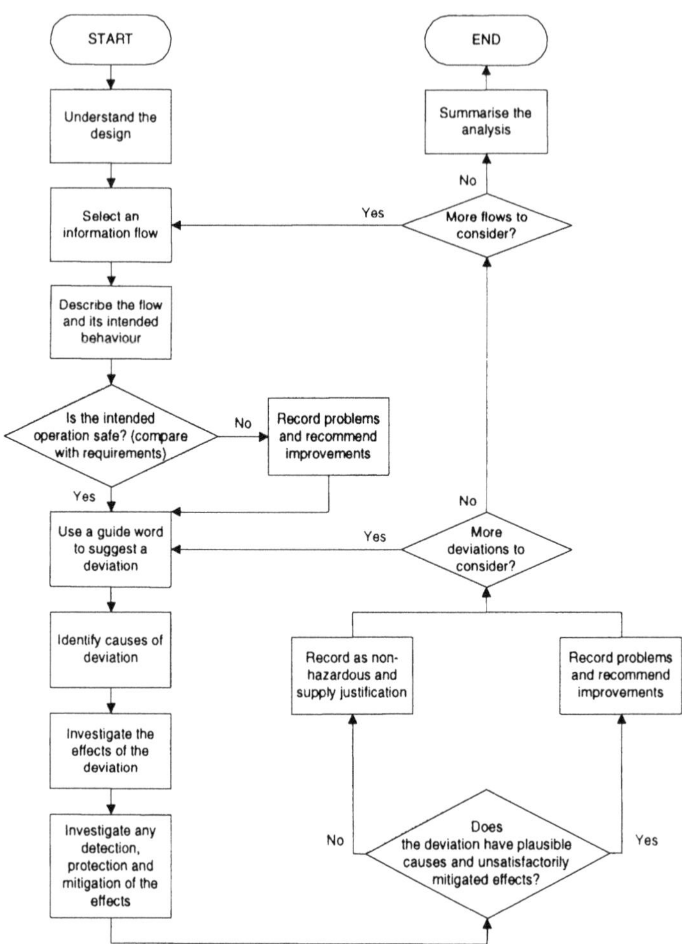

Figure 1. SHARD analysis [10]

in Section 2. An example use of the analysis of the requirements for an electronic commerce protocol is given in Section 3

2. Requirement Analysis for Security Protocol

2.1. Introduction

The aim of a requirements analysis process for security protocols is to analyse the high level requirements of the protocol to obtain the low level functional requirements of the protocol. These low level requirements can then be used in the design phase of the protocol development. This ensures that the security requirements are carefully considered at the early stages of development and the features which are built into the protocol are justified.

The analysis is split into two levels, based on the distinction between high and low level protocol requirements:

- **High Level Requirements** These state what must be achieved by the end of the protocol run. The requirements are in terms of the differences in the *knowledge* state of the principals between the start and end of the protocol session, they indicate what the principals should and should not know. They are equivalent to the pre- and post-conditions of the protocol.

  The high level requirements can be subdivided into functional and non-functional requirements. The functional requirements indicate the functionality of the system under development; for example, "By the end of the protocol, Principal A should have received an order from Principal B". At this high level we are not interested in how this is achieved, nor what the order looks like. Non-functional requirements are more difficult to analyse, these requirements include safety, security and reliability requirements. In protocol requirements analysis, we are concerned with ensuring that protocols maintain a number of security properties, which are determined by the purpose of the protocol. We have designed this analysis method with reference to the following security properties: confidentiality; authenticity; integrity; non-repudiation; availability; timeliness; non-replicability. An example of a non-functional requirement is "The order must be kept confidential between principals A and B.".

- **Low Level Requirements** These are the low level functional requirements of the protocol and are derived from the high level functional and non-functional protocol requirements. They state details such as what each protocol message will contain, how it will be constructed, any interactions between messages, such as if a particular message component is dependent on another message, and what checks will need to be carried out on the messages. An example of a low level requirement is: "The message should contain a component (such as a timestamp) to avoid replay attacks and to ensure timeliness of messages".

The analysis explores how an implementation may fail to meet its requirements, including how the external environment can affect the protocol. This may prompt further requirements of the protocol to de-

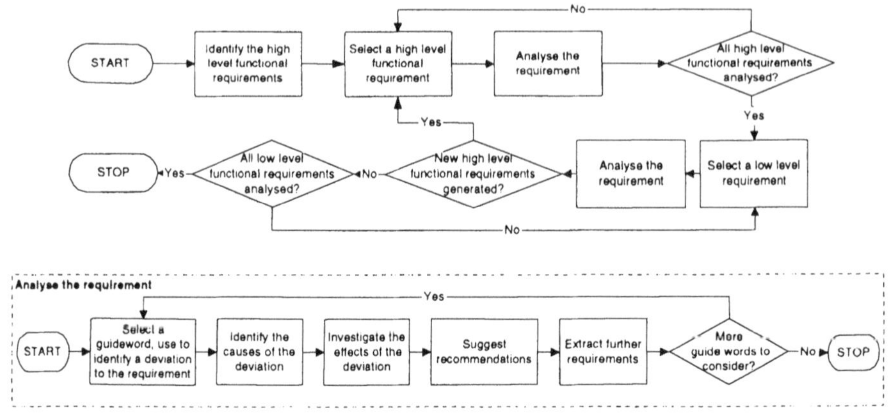

Figure 2. Hazard Analysis for Security Protocols Requirements

tect when such situations arise and also to protect or mitigate against the violations of the requirements.

Our method takes into account the differing views of the stakeholders in the protocol through the use of a team in the analysis. The team should consist of representatives of each of the stakeholders in the protocol and, ideally, someone who is familiar with attacks and flaws which are common in protocols, as well as the different verification techniques which can be used on the protocols.

2.2. Hazard Analysis for Security Protocols

The analysis process is based on the SHARD process using guide words to prompt deviations to the requirements and identifying the causes, effects, detection and mitigation mechanisms associated with these deviations.

Our analysis is carried out at the both the high and low levels of functional requirements. The analysis of the requirements will prompt further high level and low level functional requirements which will be subject to further analysis. Thus the analysis is an iterative process. It is outlined in Figure 2.

The guide words in the analysis process have been adapted to relate to message transfers, contents and checks on messages, to prompt deviations which make the requirements vulnerable to attacks which violate the required security properties. Once these vulnerabilities have been identified, measures can be taken, through the introduction of further requirements to avoid the incorporation of the vulnerabilities into the design and implementation.

Table 2 contains the guide words and interpretations used to generate the deviations to the protocol requirements, these were influenced by the SHARD guide words in Table 1. In this table we also identify the security property violations which could result from the deviations suggested by the guide words. The guide words we have selected may not be the only guide words which could be used for security protocol analysis. Further guide words can be added to the process to reflect different properties required of different types of protocol.

Some of the steps in the analysis process of Figure 2 are described in more detail below:

Identification of high level functional requirements. The high level functional requirements are elicited from the informal scenario description which details the situation for which we wish to design a protocol. By identifying the *principals*, their *actions* and the *objects* on which they act, we can extract more structured requirements which describe the scenario which contain the following: initiating principal, responding principal, action, object.

Identification of Causes. This is based on the *primary - secondary - command* rule for identifying causes in SHARD. We interpret this as:

- **Primary (P)** causes are due to the failure of the principal who carries out the service. For instance, the principal may not have sent out the message, or may have sent out an incorrect message.
- **Secondary (S)** causes are due to the failure of the medium over which communication is made or an action or event is carried out, such as the network. Cases where an intermediary party, such as an intruder causes a deviation are also classed as secondary causes.
- **Command (C)** causes are due to the failure of the command which prompts the action to be carried out. Earlier messages in a protocol session act as a command, or prompt, to the principal to send out the next message. Therefore, if an incorrect message is received then a response dependent on that message may also be incorrect.

Identification of Effects. The immediate effects of the deviations are noted. Any possible *actions* **(A)** which can be carried out by the principals as a result of the deviation are identified and the *consequences* **(C)** of these actions are identified. The actions and their consequences

GUIDE WORD	GENERIC MEANING	SECURITY PROPERTIES VIOLATED
Omission	The event does not take place.	Availability
Commission	The event which takes place is not as expected. The different types of commission: • *spurious*: a one off event. • *repetition*: a repeated event.	*Spurious:* Authenticity, Non-repudiation. *Repetition:* Authenticity Non-repudiation Non-replication.
Value	The data obtained in the event has the wrong value and this can be detected. This could be: • *total*: the data delivered in the event is totally corrupted. • *extra*: an event occurs as expected but with some unexpected extra data/behaviour. • *partial*: parts of the expected event are omitted.	Integrity, Authenticity, Non-repudiation.
Disclosure	The data in this event has been divulged to an unauthorised party.	Confidentiality.
Early	The event occurs earlier than intended. Early can be interpreted as: • *absolute*: early compared to a real-time deadline. • *relative*: early with respect to other events or communications in the system.	Authenticity, Timeliness.
Late	The event occurs later than intended. Late can be interpreted as: • *absolute*: late compared to a real-time deadline. • *relative*: late with respect to other events or communications in the system.	Authenticity, Timeliness, Availability.

Table 2. The protocol analysis guide words

must be considered because it is often a chain of events following from a deviation which results in an insecurity in the protocol. As noted in the SHARD analysis, effects may contribute to or be the causes of other deviations, therefore any dependencies should be recorded.

Recommendations. In order to protect the security of the protocol, recommendations to address the deviations are made. The recommendations are of three types:

- **Prevention (P)** Measures to prevent a potential violation of a security property, are incorporated into the design.
- **Detection (D)** Mechanisms to detect when, how and who violated the security of the protocol.
- **Reaction (R)** If we can detect when a violation has taken place then we can recover from the security violation by correction or mitigation mechanisms.

The recommendations depend upon the security properties which have been breached. In some circumstances we can react to the security violation and carry out an action to return the protocol to a secure state. However, in some cases it is impossible to recover from a security violation, such as when a confidentiality breach occurs. In such cases we must attempt to find protection mechanisms to prevent such security breaches. Similarly there are also situations in which detection of a security violation is very difficult. The choice of recommendations must be carefully considered to deal with such cases.

The recommendations can be implemented using a variety of methods. Software or hardware controls could be used to ensure that the protocol security is maintained; for example, encryption can be used to maintain integrity and confidentiality. Policies and physical controls can be used to govern the application of the protocols and information in a wider context.

Further high and low level requirements are elicited from the recommendations. These are then added to the list of requirements and are, in turn, analysed. If there are multiple recommendations to address the same problem, then design decisions about which recommendations to use will need to be taken and these should be documented.

Analysis Documentation. The analysis is documented in a table such as that in SHARD and HAZOP. The documentation table may also contain a column for recording comments arising in the course of the discussion, this is useful for recording other issues and cross references relating to other parts of the analysis or design phase. Example headers in the documentation table are: *Guide word; Deviation; Causes; Effects; Recommendations; Comments.*

3. Example Application

In this section, we provide a partial example of the use of the requirements analysis process for security protocols.

3.1. Scenario

A vendor wishes to sell goods to its customers over the internet using an electronic commerce protocol. It is envisaged that the customer will send to the vendor an order for the goods and also payment details. The vendor will then be able to obtain payment through the customer's credit card company. In return the customer will obtain the goods ordered.

3.2. Example Analysis

Identification of High Level Functional Requirements. We extracted the following high level requirements from the scenario above by identifying the principals, actions and objects and their interactions:

1 Customer sends an order for goods to the vendor.

2 Customer sends payment details to the vendor.

3 Vendor submits the payment details to the payment authority.

4 Vendor obtains payment for the goods from the payment authority.

5 Vendor distributes the goods to customer.

High Level Analysis. Table 3 contains an analysis of the high level requirement "Customer sends an order for goods to the vendor" using the hazard analysis process. In a full analysis, a similar table is produced for each of the requirements.

Omission:	No order is made.
Cause:	(P) Customer doesn't send an order. (S) Order lost by network/intruder actions.
Effect:	(A) Customer waits indefinitely for response from vendor. (C) Vendor looses an order if not detected.

Table 3. Analysis of "Customer sends order for goods to vendor"

Recommendations:	(D) Timeout on waiting for response to order so customer does not wait indefinitely. (R) Recovery session to resend order. (D) A pre-protocol exchange enables vendor to detect if an order is missing. (P) Use a reliable network.
Comments:	Network reliability is out of the scope of this protocol since we have no control over the reliability of the internet. Prevention of intruder attacks is impossible, protection should make attacks infeasible.
Commission (Spurious):	An order takes place unexpectedly.
Cause:	(P) Customer accidentally sends the order. (S) An intruder fakes an order. (S) Network fault results in spurious order.
Effect:	(A) Vendor treats order as valid and waits indefinitely for a payment message which will not take place (if payment is before delivery).
Recommendations:	(D) Timeouts on waiting for payment message so vendor doesn't wait indefinitely. (D) Order authentication. (D) Customer feedback to check order is correct. (P) A pre-protocol exchange so valid orders received by vendor are not unexpected.
Comments	As for Omission
Commission (Repetition):	An order is repeated.
Cause:	(P) Customer repeats an order intentionally. (P) Customer accidentally sends a repeat order. (S) An intruder replays the order maliciously. (S) Network fault causes message to be resent.
Effect:	(A) Vendor treats order as valid and customer receives unwanted goods. (A) Vendor rejects order and customer waits indefinitely for response from vendor.
Recommendations:	(D) Use of a fresh element (nonce or timestamp) to detect replay of an order. This allows valid repeat orders to take place. (D) Customer feedback to check order is correct.

Table 3. Analysis of "Customer sends order for goods to vendor"

Value (Total):	Order is totally corrupted
Cause:	(S) Corrupted on the network or by intruder.
Effect:	(A,C) Vendor rejects message as it is not identifiable as an order and customer waits indefinitely for response from vendor. (A) Message interpreted as an order, but not that intended by the customer.
Recommendations:	(D) Check integrity of order. (D) Customer feedback to check order is correct. (P) Avoid indefinite waiting by timing out waiting for a response to order. (R) Recovery session to resend the order (P) Use a reliable network.
Comment:	As for Omission.
Value (Extra):	Order is valid but there is some extra information with it.
Cause:	(P) Extra information added by customer. (S) Result of corruption on the network/by an intruder.
Effect:	(A) Order interpreted by vendor as an order with unwanted extra items included. (A) Order rejected by vendor and customer waits indefinitely for response from vendor.
Recommendations:	(D) Check integrity of order. (D) Customer feedback to check order is correct. (P) Avoid indefinite waiting by timing out waiting for a response to order. (R) Recovery session to resend the order. (P) Use a reliable network.
Comment:	As for Omission.
Value (Partial):	Only part of order message is received.
Cause:	(P) Customer missed off parts of order message. (S) Components of order message are lost on network/by an intruder.
Effect:	(A) Order accepted but parts of customer's order are missing. (A) Order rejected and customer waits indefinitely for response from vendor.
Recommendations:	(D) Check integrity of order. (D) Customer feedback to check order is correct.

Table 3. Analysis of "Customer sends order for goods to vendor"

	(R) Recovery session to resend the order. (P) Use a reliable network.
Comment:	As for Omission.
Disclosure:	Order is disclosed.
Cause:	(C) Order is not protected and can be read by eavesdropper on network.
Effect:	(C) Customer's privacy is violated as order is public knowledge. (C) Vendor's order details are available to everyone, including their competitors.
Recommendations:	(P) Confidentiality protection of the order.
Early:	Order is received early.
	As for Commission (spurious).
Late:	Order is received late.
Cause:	(S) Delay on network or by an intruder.
Effect:	(A) Customer waits indefinitely for vendor's response to order.
Recommendations:	(D) Inclusion of a fresh component to enable the vendor to determine if a message is late. (D) Customer times-out waiting for messages for vendor's response to avoid indefinite waiting. (R) Recovery session to resend order message. (P) Use a reliable network.
Comment	As for Omission

Table 3. Analysis of "Customer sends order for goods to vendor"

Extraction of Further Requirements from High Level Analysis. The following requirements were extracted from the analysis of the requirement "Customer sends an order for goods to vendor". In a full analysis, these are analysed in later iterations of the Hazard Analysis process.

- **High level requirements:**

  1 A recovery session should be available in case that order needs to be resent, if it is detected that order is incorrect or has not been received by the vendor.

  2 Pre-protocol exchange to ensure that vendor is alive and accepting orders and also so that vendor is able to anticipate receipt of orders.

3 Provide feedback (a confirmation of order) so customer can check that order is correct.

- **Low level requirements:**

  1 Time-outs on waiting for orders and responses to orders to avoid principals waiting indefinitely.

  2 Authentication of order messages.

  3 A fresh element in order provides uniqueness of order, giving assurance that it has been created recently and allowing orders to be repeated.

  4 Integrity checking and correction of order.

  5 Confidentiality protection of order to protect customer's privacy.

  6 Incorporation of a time component to detect if order is late.

Low Level Requirements. Table 4 contains an example of the low level analysis stage in the Hazard Analysis for Security Protocols process. This table shows the analysis of the low level requirement "A fresh element in order provides uniqueness of the order message and allows orders to be repeated." From this analysis we obtain further requirements for the protocol.

Omission:	No fresh element is included in the order message.
Cause:	(P) Not included by customer. (S) Unavailability of fresh element generator.
Effect:	(C) Vendor cannot check if order was created recently. (A) Intruder is able to replay order message.
Recommendations:	(D) Vendor checks for fresh element in order and reject order if it contains no fresh element. (P) Use of reliable fresh element generator.
Commission (spurious):	Fresh element is unexpectedly in order message.
Comment	Not applicable since message is expected to contain a fresh element.
Commission (Repetition):	A fresh element is reused in order message.
Cause:	(P) Reused by principal. (S) Element replayed by intruder/network.

Table 4. Analysis of "A fresh element is included in order message"

Effect:	(A) Rejection of message by vendor. (C) Customer does not receive goods.
Recommendations:	(D) Check fresh element and reject if repeated. (R) Recovery session to deal with invalid fresh elements.
Value (Total):	A fresh element of unexpected format is in order message.
Value (Extra):	Expected fresh element plus extra information is in order message.
Value (Partial):	Partial fresh element is in order message.
Cause:	(P) Included by customer. (S) Element in format provided by generator.
Effect:	(A) Order message is rejected by vendor. (C) Customer does not receive goods.
Recommendations:	(D) Check fresh element and reject if invalid. (R) Recovery session for cases where fresh element is invalid.
Disclosure	Fresh element is disclosed.
Cause:	(P) Not protected by principal. (S) Disclosed on network/by intruder.
Effect:	None. Public knowledge should reveal nothing.
Early	Fresh element in order message is early.
Cause:	(S) Generator dispenses fresh items too early. (S) Other messages have not yet been received.
Effect:	(C) It is known that order message has been created recently and so is valid.
Recommendations:	(P) The fresh element generator for customer and vendor should be periodically synchronised.
Late	Fresh element in order message is late.
Cause:	(P) Principal sends order message late. (S) Fresh element generator generates late. (S) Message delayed by intruder/network.
Effect:	(A) Order is rejected because it is too late.
Recommendations:	(D) Check that messages are timely/fresh and reject if late. (P) Periodic synchronisation of customer and vendor fresh element generators. (R) Recovery session in case of late messages.

Table 4. Analysis of "A fresh element is included in order message"

Extraction of Further Requirements from Low Level Analysis. The analysis of the low level requirement "A fresh element is included in the order message" identified the following requirements of the protocol:

- **High level requirements**

  1 Recovery session to deal with messages with invalid or late fresh elements.

- **Low level requirements**

  1 Periodic synchronisation of fresh element generator.

  2 Use of reliable fresh element generator.

  3 Checks to ensure fresh elements are of valid format/timely and reject if not.

  4 Checks for the fresh element in order message and rejection if no fresh element.

3.3. What has been gained from this analysis?

From this fragment of an example of a Hazard Analysis for Security Protocol requirements, we can gain insight into the intuitive steps taken by the designer. We can identify items which need to be kept confidential, checked for authenticity, integrity and freshness, recovery sessions and feedback to the principals which is required. Using this analysis process, we can trace the generation of requirements and justify the features which are built into the protocol.

In a full analysis, each of the recommendations would be justified in more detail and labelled to make it easier to trace and refer to the protocol requirements during the later development phases.

4. Conclusions

In this paper, we have described a process for the gathering and analysis of the requirements of security protocols before the actual design of the protocol. This is the traditional starting point in the software engineering life cycle. It is preferable to spend time in the early stages of the protocol development than to risk a compromise of security, when the protocol is put into use. Our approach differs from previous research into the requirements of protocols which focused on the use of requirements in the verification of protocols [14]; for example, Syverson and Meadows [15] formalised the requirements of authentication protocols and used them to verify and find attacks on the Neuman-Stubblebine protocol.

The hazard analysis approach described in this paper provides a simpler, more structured and systematic approach to deviation identification than the heuristic methods in the literature. Work on inquiry-based requirements analysis [8] relies on the use of *what-if?* questions to prompt deviations. In goal-based requirements analysis [3, 4], trivial obstacles are assigned to each goal to investigate the possible ways in which a goal may fail to complete. These obstacles are identified through the use of an extensive set of heuristics. The obstacle analysis is further elaborated through scenario analysis which examines the concrete circumstances under which goals may fail. Lamsweerde and Leitier [16] present formal and heuristic methods for obstacle identification and resolution based on temporal logic.

An advantage of the hazard analysis approach for protocol requirements over the temporal logic approach is the focus of the analysis on security features of the protocol. The temporal logic approach is very formal, requiring the gathering of the preconditions for the negation of the goal expressed in logic, these preconditions are obstacles to the goal. Some formal techniques have missed attacks due to their over abstraction of the protocols, since security attacks may be the result of the exploitation of properties which are not easily expressible in logic.

Our approach to the analysis of the requirements does not, of course, guarantee that all the attacks are avoided and secure protocols will be designed. The requirements analysis process is useful for highlighting weaknesses and flaws which have previously occurred in protocols.

Attack and threat avoidance techniques prompted by the guidelines may not be appropriate, for instance, if the recommendations would be too costly or time consuming. Consideration of the recommendations should be carefully evaluated with respect to the requirements of the protocol stakeholders. However, just being aware of potential problems which may be caused by a particular requirement is an important benefit of using the method. In such situations, if it is considered appropriate, higher level requirements may be weakened in the light of the analysis.

Our method is suitable for identifying and investigating common threats and attacks on protocols and prompting protection mechanisms against them. This method is a step forward in providing a more structured approach to the development of secure protocols and we believe that this approach to requirements analysis can be applied more widely in the field of computer security.

Acknowledgments

We would like to thank John Clark, Jonathan Moffett, colleagues in the High Integrity Systems Engineering research group and the anonymous reviewers for their helpful comments.

References

[1] Ross Anderson. How to Cheat at the Lottery (or, Massively Parallel Requirements Engineering). Invited Talk at the 15th Annual Computer Security Applications Conference, Phoenix, Arizona, December 1999.

[2] Ross J. Anderson. Why Cryptosystems Fail. *Communications of the ACM*, 37(11):32–40, 1994.

[3] Annie I. Antón. Goal-Based Requirements Analysis. In *2nd IEEE International Conference on Requirements Engineering*, pages 136–144, April 1996.

[4] Annie I. Antón. *Goal Identification and Refinement in the Specification of Software-Based Information Systems.* PhD thesis, Georgia Institute of Technology, Atlanta, June 1997.

[5] Barry W. Boehm. A Spiral Model of Software Development and Enhancement. *IEEE Computer*, pages 61–72, May 1988.

[6] CISHEC. A Guide to Hazard and Operability Studies. The Chemical Industry Safety and Health Council of the Chemical Industries Association Ltd, 1977.

[7] T. Kletz. *HAZOP and HAZAN: Identifying and Assessing Process Industry Hazards.* Institution of Chemical Engineers, third edition, 1992.

[8] Colin Potts, Kenji Takahashi, and Annie I. Antón. Inquiry-Based Requirements Analysis. *IEEE Software*, 11(2):21–32, March 1994.

[9] Roger S. Pressman. *Software Engineering: A Practitioner's Approach.* McGraw Hill, 5th edition, 2000. (European adaption by Daryl Ince).

[10] D.J. Pumfrey. The Principled Design of Computer System Safety Analyses. DPhil Thesis, University of York, 2000.

[11] Felix Redmill, Morris Chudleigh, and James Catmur. *System Safety: HAZOP and Software HAZOP.* Wiley, 1999.

[12] W. W. Royce. Managing the Development of Large Software Systems. In *Proceedings of IEEE WESCON*, pages 1–9, 1970. Reprinted in Thayer R.H.(ed.) (1988) *IEEE Tutorial on Software Engineering Project Management.*

[13] Ian Sommerville. *Software Engineering.* Addison Wesley, 6th edition, 2000.

[14] Paul Syverson and Catherine Meadows. A Logical Language for Specifying Cryptographic Protocol Requirements. In *IEEE Symposium on Research into Security and Privacy*, pages 165–177. IEEE Computer Society Press, 1993.

[15] Paul Syverson and Catherine Meadows. Formal Requirements for Key Distribution Protocols. In Alfredo De Santis, editor, *Advances in Cryptology - EUROCRYPT '94*, volume 950 of *Lecture Notes in Computer Science*, pages 320–331. Springer, May 1994.

[16] Axel van Lamsweerde and Emmanuel Letier. Handling Obstacles in Goal-Oriented Requirements Engineering. *IEEE Transactions on Software Engineering*, 26(10):978–1005, October 2000.

SECURING RMI COMMUNICATION

Vincent Naessens
K.U.Leuven, Campus Kortrijk (KULAK)
vincent.naessens@kulak.ac.be

Bart Vanhaute
K.U.Leuven, Dept. of Computer Science, DistriNet
bart.vanhaute@cs.kuleuven.ac.be

Bart De Decker
K.U.Leuven, Dept. of Computer Science, DistriNet
bart.dedecker@cs.kuleuven.ac.be

Abstract: Application programmers often have to protect their applications themselves in order to achieve secure applications. Therefore, they have to possess a lot of knowledge about security related issues. The solution to this problem is to separate the security-related modules as much as possible from the real application and transparently invoke these security modules. By doing this, the application programmer can build his distributed application without considering the security requirements.

The case study presents how to achieve transparent security in the RMI (remote method invocation) system, an API provided by Java to implement applications in a distributed environment. The presented framework is also flexible enough to support different levels of security.

Keywords: open distributed system, security framework

1. INTRODUCTION

Enterprises are increasingly dependent on their information systems to support their business activities. Compromise of these systems either —in terms of loss or inaccuracy of information or competitors gaining access to it— can be extremely costly to the enterprise. Security breaches, are becoming more frequent and varied. These may often be due to accidental misuse of the system, such as users accidentally gaining unauthorized access to information. Commercial as well as government systems may also be subject to malicious attacks (for example, to gain access to sensitive information). Distributed systems are more vulnerable to security breaches than the more traditional systems, as there are more places where the system can be attacked. Therefore, security is needed in distributed systems. This case study presents how to achieve transparent security in the RMI system.

Security protects an information system from unauthorized attempts to access information or interfere with its operation. The key security features we are concerned with are:

- **identification** and **authentication** to verify parties who they claim to be.
- **authorization** and **access control** to decide whether some party can execute some action.
- **protection of communication** between parties. This requires trust to be established between the client and the server, which involves authentication of clients to servers and authentication of servers to clients. It also requires integrity and confidentiality protection of messages in transit.
- **audit trail** of actions.

Apart from these security requirements, **administration** of security information is also needed.

In client/server applications, objects located at one host are communicating with objects running on other hosts. The key security features can be provided at two levels: at the location[1] level and at the object level. Security features provided at the location level secure communication between two hosts. This kind of security is independent of the objects communicating between these hosts. Each object can also be individually protected if security is provided at the object level. It is clear that security provided at the object level is more fine-grained than security provided at the

[1] Locations will mostly correspond with hosts; more precisely, they correspond to Java Virtual Machine instantiations.

location level. We will discuss how each of these security features can be built into the system.

The main goal of this case study is to provide a flexible and transparent security framework for the RMI system. *Flexibility* means that it must be possible to incorporate different mechanisms and services, according to the degree of security that is required. *Transparency* mains that applications are not aware of the security aspects built into the system. Hence, each of the security features should be implemented into the RMI system itself. That way, application programmers do not have to recompile their applications to work with the secured framework.

A first section briefly describes the architecture of the RMI system. The second section introduces the security components and discusses where these services should be added in the RMI system. By including these components in the RMI system itself, they are transparent with respect to the application. The third section presents a security framework for RMI that is flexible enough to support different levels of security. The next two sections discuss the transparency and the flexibility of the framework. Next, we refer to some related work in this area. The paper ends with a general conclusion.

2. THE RMI SYSTEM

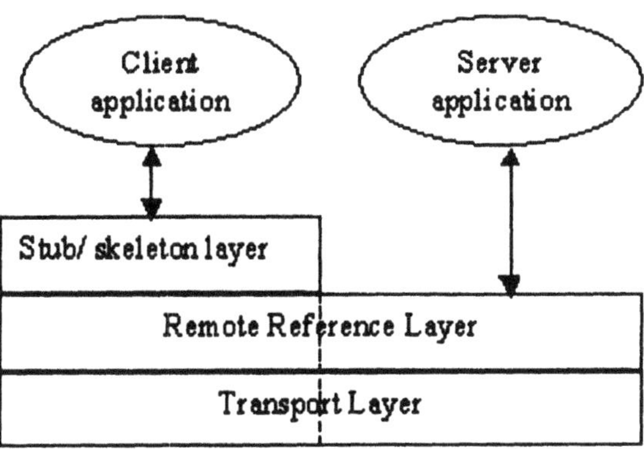

Figure 1. the RMI system

The RMI system [1] consists of three layers: the *stub/skeleton layer*, the *remote reference layer* and the *transport layer*. The application itself runs on top of this RMI system. When a client invokes an operation on a server object, a stub object passes the method to the reference layer that initiates the call. The remote references are mapped to locations. A specific reference semantics is executed at that moment depending on the implementation of the reference layer. For instance, this layer can support point-to-point calls, calls to replicated objects, etc. The remote reference layer also sets up a connection to the server side by creating a new connection or reusing an existing connection. Depending on the implementation of the transport layer, TCP [2], UDP [3] or other types of connections are supported. When a server receives information on an incoming connection, the information will be forwarded to the reference layer that executes code according to a specific semantics. Finally, the remote object executes the method and sends the result back to the client side in the same way.

3. SECURITY COMPONENTS

To achieve a secure execution environment, some security components must be added into the distributed system. The security components discussed in this paper are the association component, the authentication component, the access control component, and the audit trail component. This section shows where these four security services are added into the RMI system. By including these services in the RMI system, they are transparent to the application.

Services can be added at two levels: the location level and the object level. Services provided at the location level are executed between hosts. Information provided at that level are the IP addresses of the communicating hosts, the principals executing at each of the two hosts, etc. Services provided at the object level are executed between objects. More information is available at that level. The method name and parameters of the remote invocation are known. Moreover, an object can be running on behalf of a certain principal. An access controller at the object level can make use of this information.

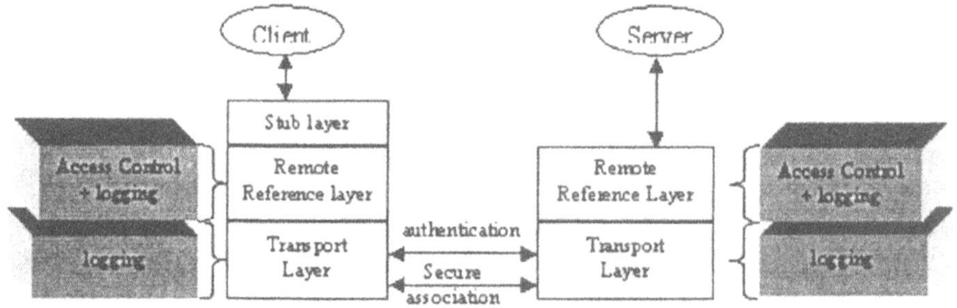

Figure 2. Security services in the RMI system.

3.1 Secure association service

Before messages are sent over the wire, a **secure association** must be established between two hosts: the client and the server. This service is provided at the location level. As a result of this phase, both parties possess a key that will be used to exchange further messages. Thus, setting up a secure association guarantees the confidentiality of the data that is exchanged between both parties. The secure RMI system performs this task after the connection is established and before the actual method invocation from the stub object to the server object takes place. This task can be fully executed at the transport layer, making use of the connection. The resulting key is also kept at the transport layer. As this service does not require any information about the objects, the same secure association can and will be reused over multiple calls between the two hosts.

3.2 Authentication service

Once a secure association is set up, an **authentication service** can be executed. Often, both parties will want to know the correct identity of the party they are dealing with, for instance as basis for authorization decisions. Alternatively, they may want to act anonymously. Authentication can be performed in a kind of handshake phase where trust is gained in the other party's identity and where security attributes are exchanged. This service can be fully performed at the transport layer, immediately after a secure association is set up. The resulting security attributes are also stored at the transport layer. Depending on the implementation, authentication is executed at the location level and/or at the object level. The presented framework only presents authentication at the location level.. This corresponds to the idea that *users* are typically controlling locations, and they are the principals we want to authenticate.

3.3 Access control service

The **access control service** (or authorisation service) gives a party the possibility to allow/disallow an action of the other party involved in the communication. In an object oriented environment, access decisions can be based on the method and the parameters that are sent to the server. This service is performed at the object level. Thus, access control must be performed at the reference layer, after the necessary information is unmarshalled and before the method will be invoked. This service can also make use of the security information that is stored at the transport layer.

3.4 Audit trail service

The **audit trail service** is responsible for logging information. Two types of logging are introduced. In the transport layer (i.e. at location level), information about the authentication procedure is logged. At the reference level (i.e. at object level), information about the authorisation and the method invocation is logged.

4. THE SECURITY FRAMEWORK

We developped a security framework for RMI that is flexible enough to support different security levels and mechanisms. By consulting a property file, the security components are loaded into the RMI system at runtime. By changing the values of this property file, other components are loaded into the system. On the one hand, objects are loaded that are responsible for holding security information. They are called **security context objects** (or security contexts). On the other hand, objects are loaded that are responsible for executing a specific security service. They are called **security service objects** (or security services). Security services can modify the information stored in the security contexts and query them to make decisions.

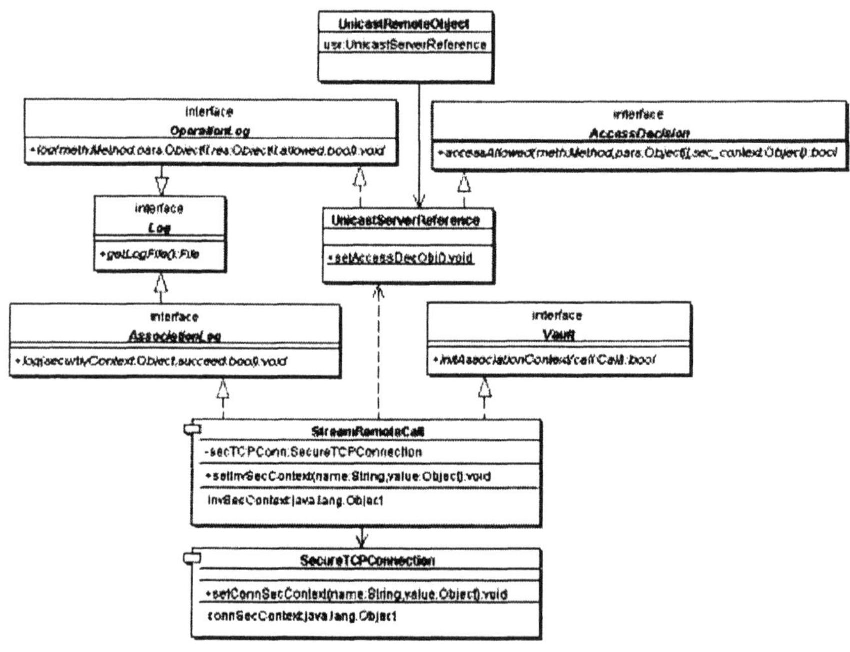

Figure 3. The security framework

4.1 Security context objects

To obtain a secure execution environment, two types of security contexts are introduced in the RMI system: a *connection security context* and *an invocation security context*. They are responsible for storing security-related data.

A **connection security context** contains security information specific for a particular connection. This context contains information exchanged during the secure association phase and the authentication phase at location level. More specifically, a connection security context can hold a session key, the time when the connection is created, the user or client that makes use of the connection, etc. Thus, every time a new connection is created, a corresponding new connection security context is initiated at the same level in the RMI system i.e. at the transport level. A connection security context disappears when the corresponding connection is closed.

An **invocation security context** holds information that is specific for a particular invocation such as the time the invocation is executed, the operation that must be executed and the parameters that belong to the operation. Thus, a new invocation security context is created each time a new call is initiated and is removed when the method call is finished. This is analogous to the first type of security context. When authentication is executed at object level, additional information is added into this context.

Remark that a connection security context can be considered a part of an invocation security context. Every invocation security context holds a pointer to a connection security context. However, the lifetime of a connection security context can be longer than the lifetime of an invocation security context. This is because the same connection can be reused during subsequent method calls.

4.2 Security service objects

Security service objects are responsible for executing some kind of security service. When a client invokes a method on a server object, a secure association is established and a particular authentication protocol is performed between the client and the server. To achieve these two tasks, a **vault object** [4] is introduced at the transport level. A vault object can perform these two tasks itself or delegate the work to an association object and an authentication object.

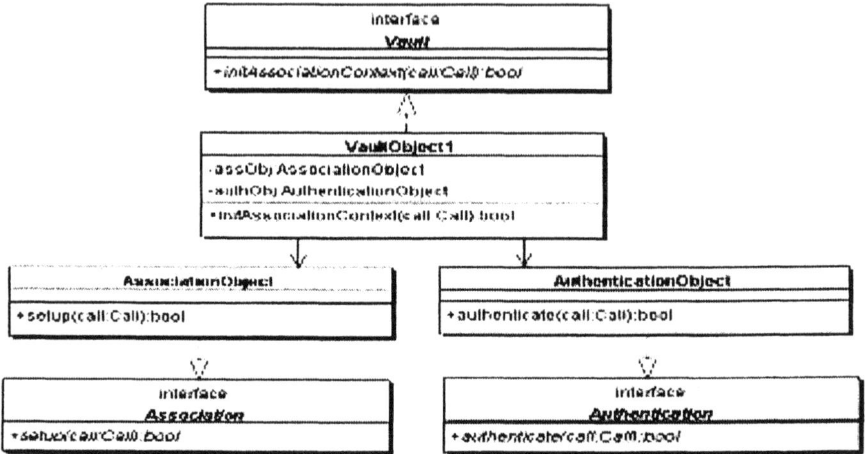

Figure 4. The Vault Object

When a call is initiated at the client side and a new unsafe connection is created, the **association object** can decide to exchange a session key with the association object on the server side. Several encryption libraries provide implementations of key agreement algorithms [5]. The resulting session key is stored at both sides in the connection security context. As a result of this step, further information can be sent in encrypted form to the connection object. In other words, encryption is done on top of a connection and therefore, it does not affect the implementation of a connection type. Moreover, if the association object sees that the connection itself is implemented to support secure communication (for instance by using SSL secure sockets), it can decide not to execute this first step. When a connection already exists, the association object can decide to update the connection security context if necessary. For instance when the time a particular key is valid, is exceeded, the vault object can ask for a new key agreement session to take place.

After this, the vault object calls an **authentication object** if authentication is not already done. Depending on its implementation, the authentication object explicitly asks the user for authentication information or makes use of credentials that are created when the user logs in on the system. These credentials are generated automatically when the user logs in on the system. It can happen that authentication is performed in several successive steps. For instance, the server side can ask for additional credentials or can conclude that the authentication data are not valid any more. In these two cases, the authentication continues. Authentication information can be sent along a secure data stream making use of the session key obtained in the previous step. The authentication information is stored in the connection security context and can later be used to make access decisions.

Access control in an object-oriented environment mostly depends on the method that must be executed and the parameters of the method call. At that point in the execution, the information must be in an unmarshalled form. Marshalling and unmarshalling happens in the remote reference layer. This information is passed to the invocation security context object. After this information is set, an **access control object** can make a decision using the information kept by the security context. At the client side, access control can be checked just before marshalling information; at the server side access control executed after unmarshalling the operation and parameters and just before the information is dispatched to the application level.

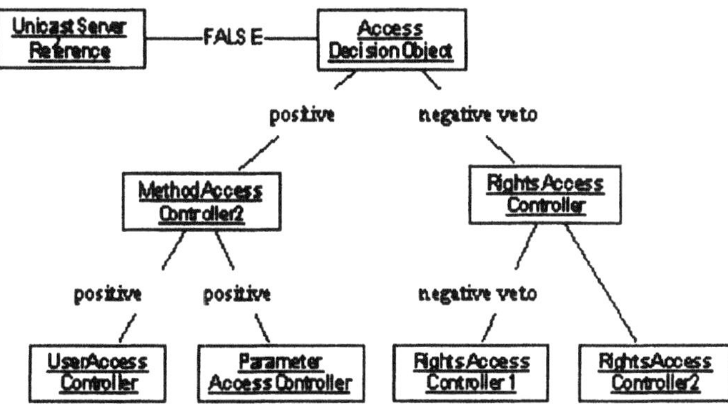

Figure 5. Access controller

To provide a flexible access control mechanism, the access control object can be implemented using the composite design pattern [6]. A tree of access controllers makes an access decision. At the leaf level, the access controllers give a negative veto or advice, or a positive veto or advice to the intermediate nodes in the tree. This information is propagated to the top level of the tree that makes a final decision. Each access controller makes a particular decision. For instance, there can be user access controllers, rights access controllers, ... These access controllers can be implemented totally independent of the actual application. To give the application the possibility to attach his own access controller to the tree, it can give a series of access controllers to the constructor of an application object. The constructor then appends the controllers to the tree in a predefined way.

Two types of log objects are introduced in the system: association log objects and operation log objects. **Association loggers** are introduced at the transport level and log information concerning the association. For instance, an association logger can save which client is trying to make a connection, if the authentication is successful, etc ... **Operation loggers** are introduced at the reference layer and log information about the operations that have to be executed or that have already been executed. For instance, an operation logger can store the method a client tries to execute, the return value of the access control decision object, the result of the method call, etc. In contrast to vault objects and access control objects, we want to provide the possibility to pick up several log objects at each level.

5. TRANSPARENCY

Because the presented security features are all built into the RMI system, it can be reused for every **application**. Access control and operation logging happens at the reference layer; setting up a secure association and logging associations happens at the transport layer. This also implies that stub objects remain the same. Therefore, the rmic compiler that generates stubs, does not have to be changed. This implies full transparency from the point of view of the application programmer.

Providing full transparency to the **end user** of the application is difficult to achieve. A secure distributed system wants the user to be authenticated at some point in the execution. Depending on the implementation of the authentication object, the user has to do it explicitly during the application runtime or the authentication object can make use of the credentials created when the user logs on the system.

From the point of view of the **administrator** of the system, one can say that he has to make a decision about which security components have to be loaded into the RMI system. He has to make a property file. The RMI system consults this property file at runtime in order to know which instances of the security components to create.

The presented framework can also be considered to be relatively transparent to the **RMI implementation** because security components are added to the system by loading security related objects and not by adapting the implementation of existing objects in the system. For instance, a typical connection implementation (UDP or TCP) does not have to be adapted because encryption is provided on top of it.

6. FLEXIBILITY

Four types of objects are introduced in the framework: security context objects, access control objects, log objects and vault objects. In turn, a vault object can call an association object and an authentication object. An appropriate interface for each of these object types is available so that the RMI system can invoke a method of an object via this interface. The property file indicates which objects to load at runtime in the system. Separating the security components from the RMI system this way provides us a flexible way of working. Although a secure RMI package can provide us with implementations of each of these objects, new implementations can be introduced as long as they implement methods of the interface in an appropriate way.

Flexibility is also needed within the proposed security components. For instance, by implementing an access decision object as a tree of access controllers, new access controllers can be added dynamically. Vault objects present a similar degree of flexibility in that way they can decide to contact an association object and an authentication object, contact one of those two types of objects or contact no other object at all according to the level of security that is preferred in the system.

7. RELATED WORK

The **Java Secure Socket Extension** (JSSE) [7] is a Java optional package that provides Secure Socket Layer (SSL) and Transport Layer Security (TLS) support for the Java 2 Platform. Using JSSE, developers can provide for the secure passage of data between a client and a server. Secure sockets can be added into the RMI system at transport level to set up a secure association. This way, they are transparent in front of application programmers. In the presented framework, the Vault object is responsible for setting up a secure association between two hosts. An implementation of that Vault object can use JSSE.

The **Java Authentication and Authorization Service** (JAAS) [7] is a framework that supplements the Java 2 platform with principal-based authentication and access control capabilities. It includes a Java implementation of the standard Pluggable Authentication Module (PAM) architecture, and provides support for user-based, group-based, or role-based access controls. These modules can also be added transparently into the presented framework. The Java Authentication Service provides authentication at object level. The framework we presented provides authentication at location level. However, we can extend the RMI security framework with authentication at object level as suggested in paragraph 3. The authorisation modules of JAAS can also be inserted into the framework in the Access Decision Object. But the security framework is flexible enough to support other types of access control. For instance, access control can also be based on the parameters and the operation that is invoked. Because Java has not specified standards for other types of authorisation, we have to make an own implementation of each of these services if that is required.

The **Common Object Services** specification (CORBASec) [4] describes security related tasks and requirements needed for CORBA. The specification is quite long and attempts to address an extremely wide range of security issues. The topic of distributed objects is complicated enough when considered on its own and it certainly does not get any simpler with the addition of security. Due to this, there are many issues that are underspecified and open to interpretation at this time, which gives scope for R&D in this area. To further extend the RMI security architecture with more advanced security services like delegation, a lot of inspiration can be found in this specification. Depending of the implementation of an ORB, different services are provided. This is similar with the flexibility of the presented RMI security framework.

The **Java Community** [8] is working on the definition of a high-level API for network security in JavaTM 2 Standard Edition RMI, covering basic security mechanisms: authentication (including delegation), confidentiality, and integrity. The main problem is that the proposals are not transparent enough towards applications. Our framework tries to achieve more transparency towards application programmers because all of the security features are built into the RMI system itself. However, the framework also enables application programmers to load their own security modules into the RMI system.

8. CONCLUSION

The presented framework gives the possibility to add different security services to the RMI system: setting up a secure association, authentication, authorisation and logging. These services are added to the RMI system in a transparent and flexible way. The implementation of the suggested objects in the framework depends on the level of security and the degree of complexity that is needed. A simple implementation can already provide a good level of security. For a more advanced implementation of each of these objects, a lot of principles suggested by security specifications of other distributed systems such as CORBA [4], can be used.

REFERENCES

1. Java™ Remote Method Invocation Specification – JDK1.2 Beta 1, October 1997
2. RMI implementation provided by SUN using TCP connections. http://java.sun.com/products/jdk/ rmi/
3. RMI implementation provided by Ninja using UDP connections. http://ninja.cs.berkeley.edu/ninja/
4. Corba Security Service Specification – november 1996.
5. IAIK: http://www.iaik.tu-graz.ac.at/
   ABA: http://aba.net.au/
6. Erich Gamma, Richard Helm, Ralph Johnson and John Vlissides, 'Design Patterns, elements of reusable objectoriënted software', Addison-Wesley 1995
7. http://www.java.sun.com/security/
8. http://java.sun.com/aboutJava/communityprocess/jsr/jsr_076_rmisecurity.html

SECURE JAVA DEVELOPMENT WITH UML

Jan Jürjens*
Computing Laboratory, University of Oxford, GB
http://www.jurjens.de/jan - jan@comlab.ox.ac.uk

Abstract Developing secure software systems is difficult and error-prone. Numerous implementations have been found vulnerable in the past; a recent example is the unauthorised access to millions of online account details at an American bank.

We aim to address this general problem in the context of development in Java. While the JDK 1.2 security architecture offers features (such as guarded objects) that provide a high degree of flexibility and the possibility to perform fine-grained access control, these features are not so easy to use correctly.

We show how to use a formal core of the Unified Modeling Language (UML), the de-facto industry-standard in object-oriented modelling, to correctly employ Java security concepts as such as signing, sealing, and guarding objects. We prove results for verification of specifications wrt. security requirements. We illustrate our approach with a (simplified) account of the development of a web-based financial application from formal specifications.

Keywords: Distributed systems security, access control, mobile code, Java security, secure software engineering, Unified Modeling Language.

1. Introduction

The need to consider security aspects in the development of many systems today is not always met by adequate knowledge on the side of the developer. This is problematic since in practice, security is compromised most often not by breaking the dedicated mechanisms (such as encryption or access control), but by exploiting weaknesses in the way they are being used [And01]. Thus security mechanisms cannot be

*Supported by the Studienstiftung des deutschen Volkes and the Computing Laboratory.

"blindly" inserted into a security-critical system, but the overall system development must take security aspects into account.

Especially dynamic access control mechanisms such as provided by Java since the JDK 1.2 security architecture [Gon99; Kar00b] in the form of GuardedObjects can be difficult to administer since it is easy to forget an access check [Gon98; BV99]. If the appropriate access controls are not performed, the security of the entire system may be compromised. Additionally, access control may be granted indirectly and unintentionally by granting access to an object containing the signature key that enables access to another object. In this work, we aim to address these problems by providing means of reasoning about the correct deployment of security mechanisms such as *signed*, *sealed* and *guarded objects* using a formal core of the widely used object-oriented design language Unified Modeling Language (UML), extending previous work [Jür01f; Jür01a].

The more general aim of this work is to use UML to encapsulate knowledge on prudent security engineering and thereby make it available to developers not specialised in security [Jür01b]. Thus the approach to use UML for security covers not just access control, but also other security functions and requirements.

Overview. After presenting some background on access control in Java in the following section, we summarise our use of UML in section 3. In Section 4 we outline the part of a design process relevant to enforcing access control in Java and give some results on verifying access control requirements. In Section 5 we illustrate our approach with the example of the development of a web-based financial application from formal specifications. We end with an account of related work, a conclusion and indication of future work. Proofs have to be omitted due to space reasons and will appear in an extended version.

2. Access control in Java

Authorisation or access control [SS94] is one of the corner-stones of computer security. The objective is to determine whether the source of a request is *authorised* to be granted the request. Distributed systems offer additional challenges: The trusted computing bases (TCBs) may be in various locations and under different controls. Communication is in presence of possible adversaries. Mobile code is employed that is possibly malicious. Further complications arise from the need for delegation (i. e. entities acting on behalf of other entities) and the fact that many security requirements are location-dependent (e.g., a user may have more rights at the office terminal than when logging in from home).

Object-oriented systems offer a very suitable framework for considering security due to their encapsulation and modularisation principles [FDR94; Var95; ND97; Gol99; Sam00].

In the JDK 1.0 security architecture, the challenges posed by mobile code were addressed by letting code from remote locations execute within a *sandbox* offering strong limitations on its execution. However, this model turned out to be too simplistic and restrictive. From JDK 1.2, a more fine-grained security architecture is employed which offers a user-definable access control, and the sophisticated concepts of signing, sealing, and guarding objects [Gon99; Kar00b].

A protection domain [SS75] is a set of entities accessible by a principal. In the JDK 1.2, permissions are granted to protection domains (which consist of classes and objects). Each object or class belongs to exactly one domain.

The system security policy set by the user (or a system administrator) is represented by a policy object instantiated from the class java.security.Policy. The security policy maps sets of running code (*protection domains*) to sets of access permissions given to the code. It is specified depending on the origin of the code (as given by a URL) and on the set of public keys corresponding to the private keys with which the code is signed.

There is a hierarchy of typed and parameterised access permissions, of which the root class is java.security.Permission and other permissions are subclassed either from the root class or one of its subclasses. Permissions consist of a target and an action. For file access permissions in the class FilePermission, the targets can be directories or files, and the actions include read, write, execute, and delete.

An access permission is granted if all callers in the current thread history belong to domains that have been granted the said permission. The history of a thread includes all classes on the current stack and also transitively inherits all classes in its parent thread when the current thread is created. This mechanism can be temporarily overridden using the static method doPrivileged().

Also, access modifiers protect sensitive fields of the JVM: For example, system classes cannot be replaced by subtyping since they are declared with access modifier final.

The sophisticated JDK 1.2 access control mechanisms are not so easy to use. The granting of permissions depends on the execution context (which however is overridden by doPrivileged(), which creates other subtleties). Sometimes, access control decisions rely on multiple threads. A thread may involve several protection domains. Thus it is not always easy to see if a given class will be granted a certain permission.

This complexity is increased by the new and rather powerful concepts of signed, sealed and guarded objects [Gon99]. A SignedObject contains the (to-be-)signed object and its signature.[1] It can be used internally as an authorisation token or to sign and serialise data or objects for storage outside the Java runtime. Nested SignedObjects can be used to construct sequences of signatures (similar to certificate chains).

Similarly, a SealedObject is an encrypted object ensuring confidentiality.

If the supplier of a resource is not in the same thread as the consumer, and the consumer thread cannot provide the access control context information, one can use a GuardedObject to protect access to the resource. The supplier of the resource creates an object representing the resource and a GuardedObject containing the resource object, and then hands the GuardedObject to the consumer. A specified Guard object incorporates checks that need to be met so that the resource object can be obtained. For this, the Guard interface contains the method checkGuard, taking an Object argument and performing the checks. To grant access the Guard objects simply returns, to deny access is throws a SecurityException. GuardedObjects are a quite powerful access control mechanism. However, their use can be difficult to administer [Gon98]. For example, access to an object may be granted indirectly (and possibly unintentionally) by giving access to another object containing the signature key for which the corresponding signature provides access to the first object.

3. Developing Secure Systems with UML

To address these issues, we extend previous work [Jür01f; Jür01a] to employ a formal core of the Unified Modeling Language (UML) [UML01], the de-facto industry standard in object-oriented modelling (an excellent introduction is given in [SP00]). We would like to ensure that the protection mechanisms that are in place do offer the required level of security. Specifically, we check the specified dynamic behaviour against expressed security policies. We do this on the level of specification (rather than the implementation level) because design mistakes can so be corrected as early as possible, and because formal reasoning is more feasible at a more abstract level.

UML consists of several kinds of diagrams describing the different views on a system. We use only a simplified fragment of UML (together with a formal semantics) to enable formal reasoning and keep

[1]Note that signing object is different from the signing of JAR files.

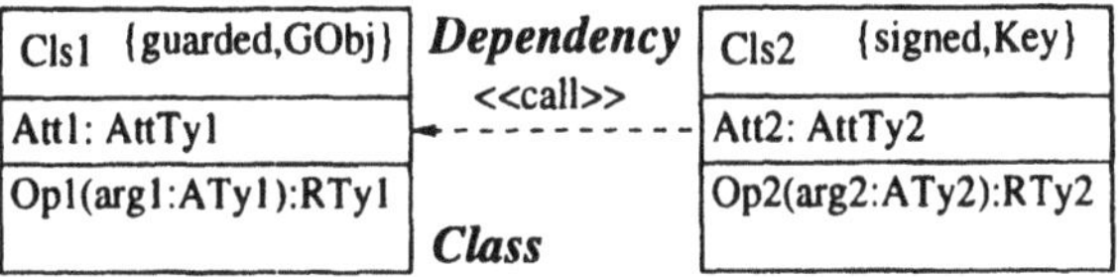

Figure 1. Class diagram

the presentation concise. We use its standard extension mechanisms to express security aspects. As a formal semantics for UML is subject of ongoing research, we use a (simplified) semantics tailored to our needs for the time being, just to illustrate our ideas. Note, however, that our approach does not rely on use of a formal semantics; in fact we aim for a tool to automatically check the considered security notions, and then these may also be explained informally (which is more accessible, but may be more prone to misunderstanding).

We use the following kinds of diagrams: class diagrams, statechart diagrams, and deployment diagrams.

We define the diagrams using their abstract syntax for conciseness and to enable formal reasoning. We also give the concrete syntax (in a way that the translation between the two should be apparent).

3.1. Class Diagrams

Using class diagrams we can model which objects are signed or sealed with which keys, and which are guarded by which Guard objects.

An *attribute specification* $A = (\mathsf{att\_name}, \mathsf{att\_type}, \mathsf{init\_value})$ is given by a name att_name, a type att_type and an initial value init_value.

An *operation specification* $O = (\mathsf{op\_name}, \mathsf{Arguments}, \mathsf{op\_type})$ is given by a name op_name, a set of Arguments and the type op_type of the return value. The set of arguments may be empty and the return type may be the empty type $\emptyset$ denoting absence of a return value. An *argument* $A = (\mathsf{arg\_name}, \mathsf{arg\_type})$ is given by its name arg_name and its type arg_type.

A *class model* $C = (\mathsf{class\_name}, (\mathsf{tag}, \mathsf{value}), \mathsf{AttSpecs}, \mathsf{OpSpecs}, \mathsf{State})$ is given by a name class_name, an optional $(\mathsf{tag}, \mathsf{value})$ pair (written in curly brackets), a set of attribute specifications AttSpecs, a set of operation specifications OpSpecs and a statechart diagram State giving the object behaviour. The tag may be either of signed, sealed or guarded (indicating a signed, sealed or guarded object), and the value is either the public key corresponding to the private key with which the object was signed or sealed, or it is the name of the corresponding Guard object.

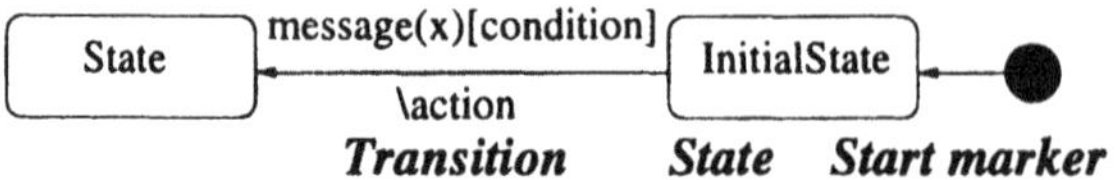

Figure 2. Statechart Diagrams

A *class diagram* $D = (\mathsf{Cls}, \mathsf{Dependencies})$ is given by a set Cls of class models and a set of Dependencies. A *dependency* is a tuple (client, supplier, stereotype) consisting of class names client and supplier and a label (called stereotype) indicating the kind of dependency (e.g. «*call*»).

3.2. Statechart diagrams

We use statechart diagrams to specify the behaviour of objects, in particular of the Guards.

We fix a set Var of (typed) variables $x, z, y, \ldots$. We define the notion of a statechart diagram for a given class model C: A *statechart diagram* $S = (\mathsf{States}, \mathsf{init\_state}, \mathsf{Transitions})$ is given by a set of States (that includes the initial state init_state) and a set of Transitions. (In the concrete syntax, the initial state is signified with a start marker.)

A *statechart transition* $t = (\mathsf{source}, \mathsf{event}, \mathsf{condition}, \mathsf{Actions}, \mathsf{target})$ has a source state, an event, a condition, a list of Actions and a target state. An *event* is the name of an operation with a list of distinct variables as arguments (e.g. $\mathsf{op}(x, y, z)$). Let the set Assignments consist of all partial functions that assign to each variable and each attribute of the class C a value of its type. A *condition*[2] is a function $g : \mathsf{Assigments} \rightarrow \mathsf{Bool}$ evaluating each assignment to a boolean value. We write it as a sequence of Boolean propositions with variables and attribute names that is interpreted as their conjunction; conditions are written in square brackets. An *action* can be either to assign a value v to an attribute a (written $a := v$), to call an operation op with values $v_1, \ldots, v_n$ (written $\mathsf{op}(v_1, \ldots, v_n)$), to return values $v_1, \ldots, v_n$ as a response to an earlier call of the operation op (written $\mathsf{return}_{\mathsf{op}}(v_1, \ldots, v_n)$), or to throw an exception. In each case, the values can be constants, variables or attributes. In the concrete syntax, actions are preceded by a backslash.

3.3. Deployment diagrams

Deployment diagrams describing the physical layer of a system are security-relevant in so far as they give the locations of the different components of the system (used in the access permissions) and they give

[2]We do not use the UML term *guard* here to avoid confusion with guard objects.

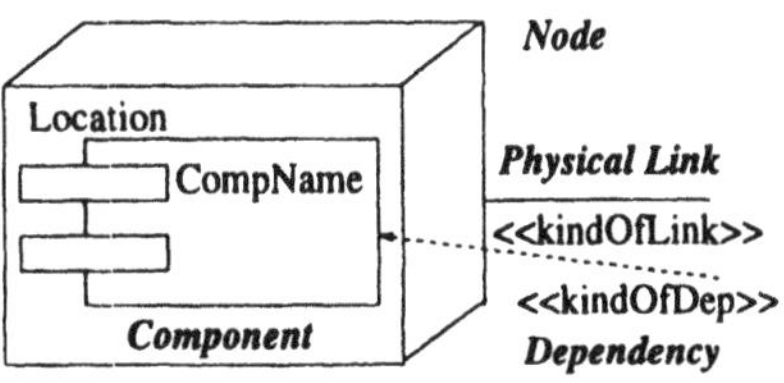

Figure 3. Deployment diagram

information on kinds of the communication links between different components, inducing threat scenarios wrt. the physical security.

A system *node* $N = (\mathsf{location}, \mathsf{Components})$ is given by its location (e.g. a URL or "local system") and a set of contained Components.

A *deployment diagram* $D = (\mathsf{Nodes}, \mathsf{Links}, \mathsf{Dependencies})$ is given by a set of Nodes, a set of communication Links between nodes and a set of logical Dependencies between components. A *link* $l = (\mathsf{nds}, \mathsf{stereo})$ consists of a two-element set nds of nodes being linked and a label (called *stereotype*) indicating the kind of the link (e.g. «*Internet*»). Here a *dependency* is a tuple $(\mathsf{client}, \mathsf{supplier}, \mathsf{interface}, \mathsf{tag})$ consisting of components client and supplier and a label (called tag) indicating the kind of dependency (e.g. $\{\mathsf{rmi}\}$).

4. Design process

We sketch the part of a design process for secure systems using UML that is concerned with access control enforcement using guarded objects.

(1) Formulate the permission sets for access control for sensitive objects.

(2) Use statecharts to specify Guard objects that enforce appropriate access control checks.

(3) Verify that the Guard objects protect the sensitive objects sufficiently by showing that they only grant access implied by the security requirements.

(4) Ensure that the access control mechanisms are consistent with the functionality required by the system by showing that the other objects may perform their intended behaviour.

(5) Verify that mobile objects are sufficiently protected by considering the threat scenario arising from the physical layer given in the deployment diagram.

Here the access control requirements in step (1) can be of the following form:[3]

- origin of requesting object (based on URL)
- signatures of requesting object
- external variables (such as time of day etc.).

In Section 5 we sketch a formal verification of a specification following these steps. They enforce the following two requirements.

Security requirement: Check that the access control requirements are strong enough to prevent unauthorised influence, given the threat scenario arising from the physical layer.

Functionality requirement: Check that the access control requirements formulated are not overly restrictive, denying legitimate access from other components of the specification.

The functionality requirement is important since it is not always easy to see if stated security requirements are at all implementable. If their inconsistency is only noticed during implementation then, firstly, resources are wasted since work has to be redone. Secondly, most likely security will be degraded in order to reduce this extra work.

4.1. Verification

In this subsection, we sketch results to be applied in the above approach. The idea is to verify security properties by linking the different views on a system given by the various kinds of diagrams. We convey our ideas using a simplified semantics for UML statechart diagrams.

Any statechart diagram S defines a function $[\![S]\!]$ from sequences of input events to sets of sequences of output actions, each possibly with arguments, often involving use of cryptographic operations (as detailed in [Jür01f]). We say that S *may eventually output* a value v if there exists a sequence $\vec{e}$ of input events and a sequence $\vec{a} \in [\![S]\!](\vec{e})$ of corresponding output actions such that v is output by one of the actions in $\vec{a}$ (in cleartext) [Jür01e].

The following definition uses the notion of an *adversary* from [Jür01e], which is a function from sequences of output actions of the statechart S to sequences of input events of S that captures the capabilities of an

[3] In future work we intend to formalise these requirements using an abstract security policy specification language, enabling automatic generation of the corresponding guard object specifications.

adversary intercepting the «*Internet*» communication links between S and the other objects (the exact definition of "adversary", "without prior knowledge" and of the composition $\otimes$ of the statechart interpretation $[\![S]\!]$ with the adversary A can be found in [Jür01e]).

Definition 1 A statechart diagram S *preserves the secrecy* of a value K if there is no adversary A (eavesdropping on the «*Internet*» links) without prior knowledge of K such that $[\![S]\!] \otimes A$ may eventually output K.

This definition is extended to system components by composing the functions arising from the statechart diagrams specifying the objects of a given component.

Intuitively, then, a system component C preserves the secrecy of K if no adversary can find out K in interaction with the system modeled by C, following the approach of Dolev and Yao (1983), cf. [Aba00; Jür01e].

The following result is applied within the approach of subsection 4 to the UML specification of a security-critical system (for a proof of this as well as the following results cf. [Jür01d]).

Theorem 1 *Suppose that the access to a certain resource is according to the* Guard *object specifications granted only to objects signed with a key K. Suppose all components preserve the security of K. Then only objects signed with K according to the specification will be granted access to the resource.*

Before coming to the main example in the next section, we give a short example to point out that the kind of weaknesses in Java security access control can be quite subtle (rather than just mistakingly sending out secret keys or forgetting to set access rules):

Example. The statechart in Figure 4 describes the behaviour of a guard object grd enforcing a slightly more complicated access control policy. The idea is that an entity named req may establish a shared key K_M in order to submit keys K_S protected by K_M such that objects signed with K_S should be granted access to the guarded object. Here we assume that the keys K_S may be updated frequently, so that it is more efficient to use the symmetric key K_M to protect K_S (rather than the public key associated with grd). The identity of req is taken as given and is bound to a public key in the certificate cert signed with the key K_C of a certification authority (assuming RSA-type encryption and signing). On request cert(), the guard object sends out a self-signed certificate certifying its public key K. The object req sends back the symmetric key

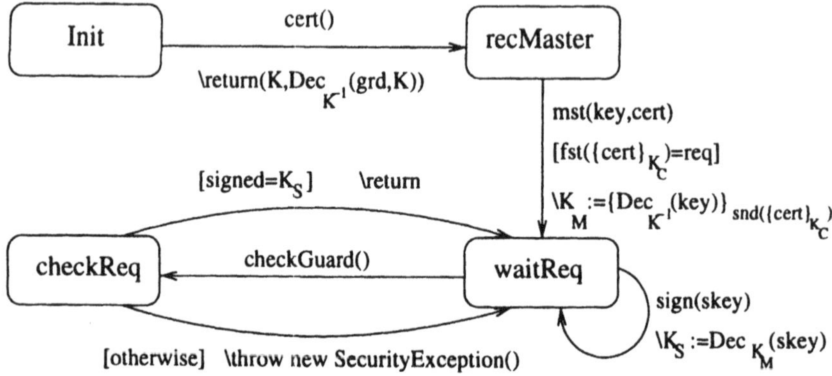

Figure 4. Statechart Example

K_M signed with its private key (corresponding to the public key in cert) and encrypted under K, together with the certificate cert (the functions fst resp. snd applied to a pair returns its first resp. second component). The guard object can receive the signature key K_S encrypted under K_M and will then grant access to those objects signed by K_S.

Thus a typical message exchange to establish K_S may look like the following:

grd $\xleftarrow{\text{cert()}}$ req

grd $\xrightarrow{\text{return}(K, \mathcal{D}ec_{K^{-1}}(grd, K))}$ req

grd $\xleftarrow{\text{mst}(\{\mathcal{D}ec_{K_r^{-1}}(K_M)\}_K, \mathcal{D}ec_{K_C^{-1}}(req, K_r))}$ req

grd $\xleftarrow{\text{sign}(\{K_S\}_{K_M})}$ req

Unfortunately, this access control mechanism contains a flaw: An adversary A intercepting the communication between req and grd (and modifying the exchanged values) can find out K_M and thus make grd accept a key K_S chosen by A. The critical part of the message exchange corresponding to this attack is as follows:

grd $\xrightarrow{\text{return}(K, \mathcal{D}ec_{K^{-1}}(grd, K))}$ A $\xrightarrow{\text{return}(K_A, \mathcal{D}ec_{K_A^{-1}}(grd, K_A))}$ req

grd $\xleftarrow{\text{mst}(\{\mathcal{D}ec_{K_r^{-1}}(K_M)\}_K, \mathcal{D}ec_{K_C^{-1}}(req, K_r))}$ A $\xleftarrow{\text{mst}(\{\mathcal{D}ec_{K_r^{-1}}(K_M)\}_{K_A}, \mathcal{D}ec_{K_C^{-1}}(reqK_r))}$ req

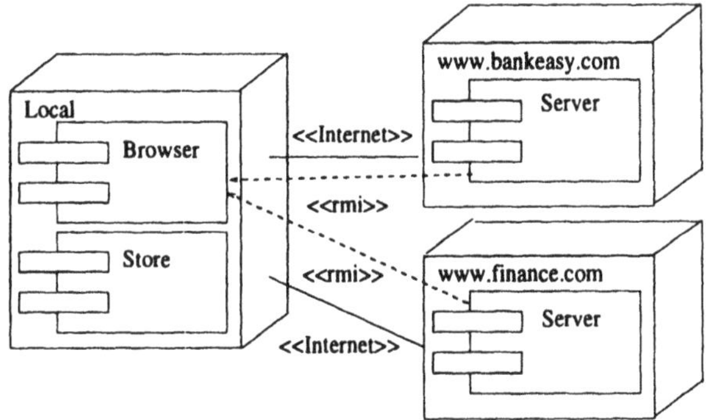

Figure 5. Deployment diagram

Here the theorem above does not apply since the security of the signing key K_S is violated (in a subtle way). With our approach one can exhibit subtle flaws like this (in this case, one would notice the flaw e.g. when trying to show formally that the assumptions of the above theorem are fulfilled). – This example is quite realistic; in fact it is derived from a published protocol which was found to be flawed in [Jür01e] (cf. there for details).

5. Example Financial Application

We illustrate our approach with the example of a web-based financial application. The example was chosen to be tractable enough given the space restrictions but still realistic in that it points out some typical issues when considering access control for web-based e-commerce applications (namely to have several entities – service-providers and customers – interacting with each other while granting the other parties a limited amount of trust and by enforcing this using credentials).

We first describe the physical layer of the application in a UML diagram and state its security requirements. We show in UML diagrams how to employ GuardedObjects to enforce these security requirements. We prove that the specification given by the UML diagrams is secure by showing that it does not grant any access not implied by the security requirements. We end the section by giving supplementary results regarding consistency of the security requirements.

Two (fictional) institutions offer services over the Internet to local users: an Internet bank, Bankeasy, and a financial advisor, Finance. The physical layer is thus given in Figure 5.

To make use of these services, a local client needs to grant the applets from the respective sites certain privileges.

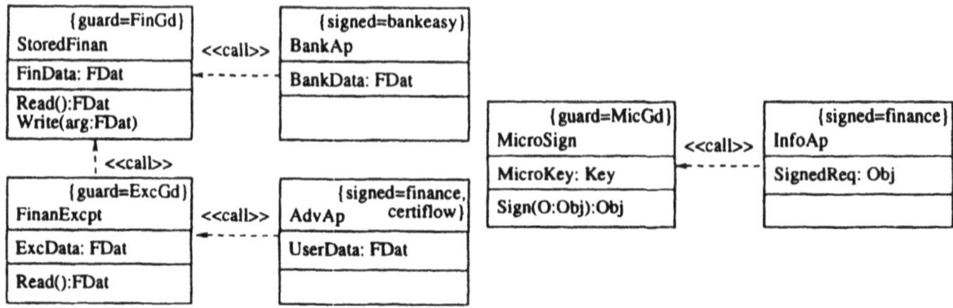

Figure 6. Class diagram

(1) Applets that originate at and are signed by the bank can read and write the financial data stored in the local database, but only between 1 pm and 2 pm (when the user usually manages her bank account).

(2) Applets from (and signed by) the financial advisor may read an excerpt of the local financial data created for this purpose. Since this information should only be used locally, they additionally have to be signed[4] by a certification company, CertiFlow, certifying that they do not leak out information via covert channels.

(3) Applets originating at and signed by the financial advisor may use the micropayment signature key of the local user (to purchase stock rate information on behalf of the user), but this access should only be granted five times a week.

Financial data sent over the Internet is signed and sealed to ensure integrity and confidentiality. Access to the local financial data is realised using GuardedObjects. Thus the relevant part of the class diagram is given in Figure 6.

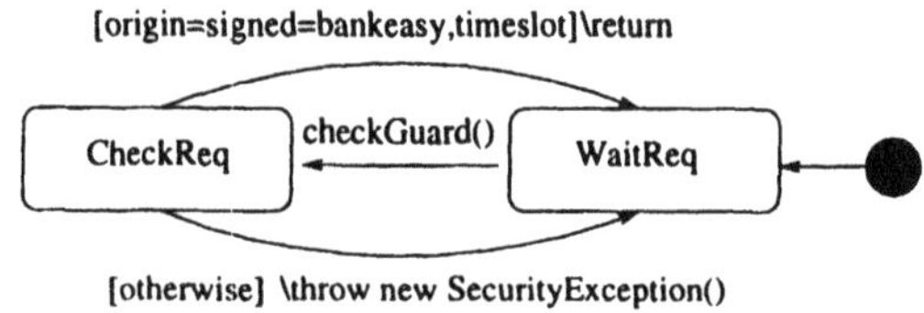

Figure 7. Statechart FinGd

[4]Here we assume that SignedObject is subclassed to allow multiple signatures on the same object [Gon99].

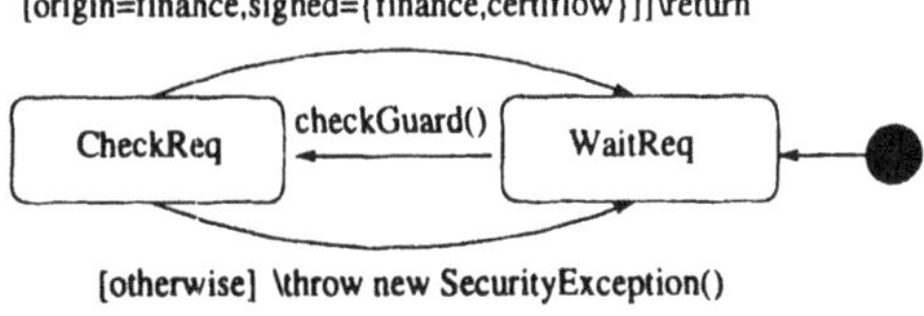

Figure 8. Statechart ExcGd

As specified in the class diagram, the access controls are realised by the Guard objects FinGd, ExpGd and MicGd, whose behaviour is specified in Figures 7, 8 and 9 (we assume that the condition timeslot is fulfilled if and only if the time is between 1pm and 2pm, that the condition weeklimit is fulfilled if and only if the access to the micropayment key has been granted less than five times in the current calendar week, and that the method incThisWeek increments the relevant counter).

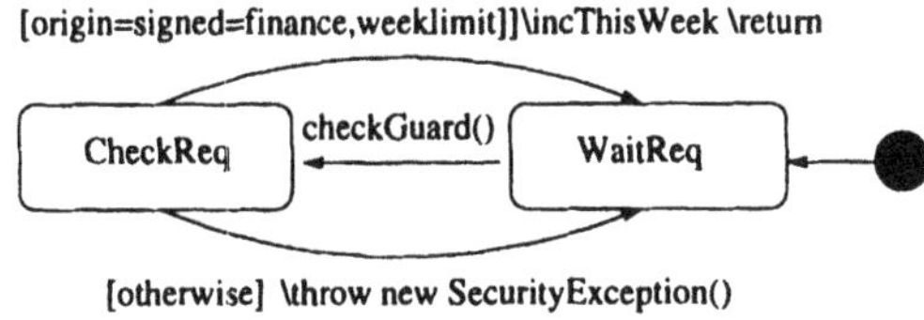

Figure 9. Statechart MicGd

Now according to step (3) in Section 4, we prove that the specification given by UML diagrams is secure in the following sense.

Theorem 2 *The specification given by UML diagrams for the guard objects does not grant any permissions not implied by the access permission requirements given in (1)-(3).*

Regarding step (4) in Section 4, we exemplarily prove that InfoAp can purchase the article on behalf of the user, as intended.

Theorem 3 *Suppose all applets in the current execution context originate from and are signed by Finance, and that use of the micropayment key is requested, which has happened less than five times before in the current week. Then the current applet is permitted to purchase articles on behalf of the user.*

Finally, following (5) in Section 4, the mobile objects are sufficiently protected since all objects sent over the Internet were required to be signed and sealed (a more detailed discussion has to be omitted).

6. Related Work

In [Jür01f; Jür01a] we considered how to model various aspects of general systems security (including multi-level security, secure information flow and security protocols) with UML. [Jür01c] applies UML to reason about audit-security in a smart-card based payment scheme and [Jür01b] shows how to use UML to enforce general principles of secure systems design from [SS75]. There seems to be little other systematic work yet in applying UML to security.

Java 2 security and in particular the advanced topics of signed, sealed and guarded objects is explained in [Gon99]. There has also been some work giving formal reference models for Java 2 access control mechanisms, thus clarifying possible ambiguities in the informal accounts and enabling proof of compiler conformance to the specification [KG98; WF98; Kar00b] (but without considering signed, sealed or guarded objects). To our knowledge, the use of signed, sealed or guarded objects in JDK 1.2 has not previously been considered in a formal model.

[HKK00] introduces higher-level abstractions for Java security policy rules, simplifies security management and gives additional functionality. General Java security is considered e.g. in [GAS99].

There has been extensive work regarding formal models for security, mostly about security protocols (for an overview cf. [GSG99; RSG+01]). A logic for access control was introduced in [ABLP93].

7. Conclusion and Future Work

To summarise, we used a core of UML, the industry standard in object-oriented modelling, to specify and reason about access control in distributed Java-based systems. We have concentrated on advanced JDK 1.2 access control mechanisms such as signing, sealing and guarding objects. We show how to specify security requirements and to prove that modelled access control mechanisms such as guarded objects meet their goals and that these mechanisms are consistent with the overall functionality required from the system.

In conclusion, it seems that our approach is both worthwhile and feasible:

- Using the JDK 1.2 access control mechanisms can be rather complicated in practice (especially when indirect access permissions using authorisation tokens are employed), thus providing support for correct specification of the relevant mechanisms in the context of a widely used specification as UML seems quite useful.

- In this paper, we could only illustrate our approach using a rather simple example. However, UML allows a high degree of abstraction in modelling systems. So we expect the approach to scale up rather well. This is currently validated in practice in a Master's thesis developing an Internet-based auction system [Mea01].

A further benefit is that by using a widely accepted notation, our approach to secure Java development can be integrated with other work on secure systems using UML (e.g. on electronic purse systems [Jür01c]).

As to the limitations of this first step in this direction of research, our account remains relatively abstract for space restrictions and conciseness of presentation. As a next step, one should consider more details of Java security, such as the use of access modifiers (private, final,...), the doPrivileged() method and the implies() method. Also, an extension to JAAS [LGK+99; Kar00b] is planned.

Work in progress aims to provide tool support to validate UML specifications of access control guards against security requirements, building on work in [CCR01].

Regarding future work, it would be very useful to have a way to generate the correct behaviour specification of guard objects in statechart diagrams automatically from the (formalised) security requirements. Also, it would be interesting to try to extend our approach to the extension of the Java security architecture proposed in [HKK00]. We intend to address CORBA security (cf. e.g. [VH96; Kar00a]) in a similar way.

Acknowledgments

The idea for the line of work using UML for security arose when doing security consulting for a project during a research visit with M. Abadi at Bell Labs (Lucent Tech.), Palo Alto, whose hospitality is gratefully acknowledged. This work benefitted from discussions at the summer school "Foundations of Security Analysis and Design 2000" (in particular, Li Gong suggested to apply the UML-based approach to security to guarded objects) and the Dagstuhl seminar "Security through Analysis and Verification" (in particular with D. Gollmann and B. Pfitzmann). The work was presented in two talks at the Computing Laboratory at the University of Oxford. Comments from S. Abramsky, C. Crichton and G. Lowe are gratefully acknowledged. Finally, comments by the anonymous referees have been very helpful.

References

[Aba00] M. Abadi. Security protocols and their properties. In F. Bauer and R. Steinbrueggen, editors, *Foundations of Secure Computation.* IOS Press, 2000.

[ABLP93] M. Abadi, Michael Burrows, Butler Lampson, and Gordon Plotkin. A calculus for access control in distributed systems. *ACM Transactions on Programming Languages and Systems*, 15(4):706–734, 1993.

[And01] R. Anderson. *Security Engineering: A Guide to Building Dependable Distributed Systems.* Wiley, 2001.

[BV99] B. Bokowski and J. Vitek. Confined types. In *14th Annual ACM SIGPLAN Conference on ObjectOriented Programming Systems, Languages, and Applications (OOPSLA'99)*, 1999.

[CCR01] R. Campo, A. Cavarra, and E. Riccobene. Simulating UML state machines. In E. Börger and U. Glässer, editors, *ASM'2001*, LNCS. Springer-Verlag, 2001. To be published.

[FDR94] J. C. Fabre, Y. Deswarte, and B. Randell. Designing secure and reliable applications using fragmentation-redundancy-scattering: an object-oriented approach. In *PDCS 2: Open Conference*, pages 343–362, Newcastle-upon-Tyne, 1994. Dept of Computing Science, University of Newcastle, NE1 7RU, UK.

[GAS99] Stefanos Gritzalis, George Aggelis, and Diomidis Spinellis. Architectures for secure portable executable content. *Internet Research*, 9(1):16–24, 1999.

[Gol99] D. Gollmann. *Computer Security.* J. Wiley, 1999.

[Gon98] Li Gong. Java^{TM} Security Architecture (JDK1.2). http://java.sun.com/products/jdk/1.2/docs/guide/security/spec/security-spec.doc.html, October 2 1998.

[Gon99] Li Gong. *Inside Java 2 Platform Security - Architecture, API Design, and Implementation.* Addison-Wesley, 1999.

[GSG99] Stefanos Gritzalis, Diomidis Spinellis, and Panagiotis Georgiadis. Security protocols over open networks and distributed systems: Formal methods for their analysis, design, and verification. *Computer Communications Journal*, 22(8):695–707, 1999.

[HKK00] Manfred Hauswirth, Clemens Kerer, and Roman Kurmanowytsch. A secure execution framework for Java. In *ACM conference on Computer and communications security*, 2000.

[Huß01] H. Hußmann, editor. *Fundamental Approaches to Software Engineering (FASE/ETAPS, International Conference)*, volume 2029 of *LNCS.* Springer-Verlag, 2001.

[Jür01a] Jan Jürjens. Developing secure systems with UMLsec — from business processes to implementation. In *VIS 2001.* Vieweg-Verlag, 2001. To appear.

[Jür01b] Jan Jürjens. Encapsulating rules of prudent security engineering. In *International Workshop on Security Protocols*, LNCS. Springer-Verlag, 2001. To be published.

[Jür01c] Jan Jürjens. Modelling audit security for smart-card payment schemes with UMLsec. In M. Dupuy and P. Paradinas, editors, *Trusted Information: The New Decade Challenge*, pages 93–108. International Federation for Information Processing (IFIP), Kluwer Academic Publishers, 2001. Proceedings of SEC 2001 – 16th International Conference on Information Security.

[Jür01d] Jan Jürjens. *Principles for Secure Systems Design*. PhD thesis, Oxford University Computing Laboratory, 2001. In preparation.

[Jür01e] Jan Jürjens. Secrecy-preserving refinement. In *Formal Methods Europe (International Symposium)*, volume 2021 of *LNCS*, pages 135–152. Springer-Verlag, 2001.

[Jür01f] Jan Jürjens. Towards development of secure systems using UMLsec. In Hußmann [Huß01], pages 187–200. Also OUCL TR-9-00 (Nov. 2000), http://web.comlab.ox.ac.uk/oucl/publications/tr/tr-9-00.html.

[Kar00a] G. Karjoth. Authorization in CORBA security. *Journal of Computer Security*, 8(2,3):89–108, 2000.

[Kar00b] G. Karjoth. Java and mobile code security – an operational semantics of Java 2 access control. In *IEEE Computer Security Foundations Workshop*, 2000.

[KG98] L. Kassab and S. Greenwald. Towards formalizing the Java Security Architecture in JDK 1.2. In *European Symposium on Research in Computer Security (ESORICS)*, LNCS. Springer-Verlag, 1998.

[LGK+99] C. Lai, L. Gong, L. Koved, A. Nadalin, and R. Schemers. User authentication and authorization in the Java platform. In *IEEE Annual Computer Security Applications Conference*, 1999.

[Mea01] W. Measor. Secure byzantine agreement – design, implementation and verification. Master's thesis, Oxford University Computing Laboratory, 2001.

[ND97] V. Nicomette and Y. Deswarte. An authorization scheme for distributed object systems. In *IEEE Symposium on Security and Privacy*, 1997.

[RSG+01] P. Ryan, S. Schneider, M. Goldsmith, G. Lowe, and B. Roscoe. *The Modelling and Analysis of Security Protocols: the CSP Approach*. Addison-Wesley, 2001.

[Sam00] P. Samarati. Access control: Policies, models, architectures, and mechanisms. Lecture Notes, 2000.

[SP00] P. Stevens and R. Pooley. *Using UML*. Addison-Wesley, 2000.

[SS75] J. Saltzer and M. Schroeder. The protection of information in computer systems. *Proceedings of the IEEE*, 63(9):1278–1308, September 1975.

[SS94] R. Sandhu and P. Samarati. Access control: Principles and practice. *IEEE Communications*, 32(9), 1994.

[UML01] UML Revision Task Force. OMG UML Specification v. 1.4 (draft). OMG Document ad/01-02-14. Available at http://www.omg.org/uml, February 2001.

[Var95] V. Varadharajan. Distributed object system security. In H.P. Eloff and S.H. von Solms, editors, *Information Security - the next Decade*, pages 305–321. Chapman & Hall, 1995.

[VH96] V. Varadharajan and T. Hardjono. Security model for distributed object framework and its applicability to CORBA. In *12th International Information Security Conference IFIP SEC'96*, 1996.

[WF98] D. Wallach and E. Felten. Understanding Java Stack Inspection. In *IEEE Security and Privacy*, 1998.

SECURITY THROUGH ASPECT-ORIENTED PROGRAMMING

Bart De Win, Bart Vanhaute and Bart De Decker
K.U.Leuven, Departement of Computer Science
Celestijnenlaan 200A, B-3001 Heverlee, Belgium
bart.dewin@cs.kuleuven.ac.be

Abstract Since many applications are too complex to be solved ad hoc, mechanisms are being developed to deal with different concerns separately. An interesting case of this separation is security. The implementation of security mechanisms often interacts or even interferes with the core functionality of the application. This results in tangled, unmanageable code with a higher risk of security bugs.

Aspect-oriented programming promises to tackle this problem by offering several abstractions that help to reason about and specify the concerns one at a time. In this paper we make use of this approach to introduce security into an application. By means of the example of access control, we investigate how well the state of the art in aspect-oriented programming can deal with the separation of security from an application. We also discuss the benefits and drawbacks of this approach, and how it relates to similar techniques.

Keywords: aspect-oriented programming, security, separation of concerns

1. Introduction

In the open world of the Internet it is very important to use secure applications, servers and operating systems in order to avoid losing valuable assets. According to different sources (e.g. CERT [cer, 2001]) updating and patching these systems to fix security holes is necessary frequently. The fact is that writing a *secure* application in an open, distributed environment is a far from straightforward task. There are several reasons why this is so hard to achieve.

First, securing an application is a very complex matter and requires a thorough understanding of what can go wrong and might be exploited. An average application programmer has not enough expertise in this area to know the exact requirements for his specific case. Moreover, compared to the core functionality of an application, security is often considered as less important and as such it is only added afterwards. This results in overlooking some details which might introduce security holes. Furthermore, in the supposition that the security requirements are clearly understood and known in advance, it is necessary to implement them in the correct way. Several examples exist (e.g. the Netscape random generator bug [Demailly, 1996]) where a small weakness or bug in security related code could bring down the security of the whole application.

A second important reason why applications are difficult to secure is the structural difference between an application and the required security solution. Confidentiality for instance requires sealing and unsealing of sensitive information. Although they are logically joint and in fact very similar, they are typically separated in the implementation. Contrary to the application logic, security deals with principals that use particular services of the application and by doing so exchange sensitive information. Clearly, this information is scattered throughout the functionality of the application. This structural mismatch often leads to duplication of security code over different locations. Management of software is complicated considerably in this way, which unfortunately often introduces security problems.

To solve these problems, this paper uses aspect-oriented programming to implement security. The separation of concerns offered by this technique allows a programmer to only focus on the core functionality of his application. At the same time, a security engineer can analyze the security requirements and add these to the application without difficulty. As an extra advantage, the security requirement implementations can be reused for other applications when properly designed. Moreover, since the security related aspects are separated from the actual application, there is no reason to fear losing the overall security picture and as such forgetting important issues.

The structure of this paper is as follows. We will first give a short introduction to AspectJ, an aspect-oriented programming language for Java, and explain how this can be used to secure an application by means of a concrete example. This mechanism will be generalized in order to construct a framework of security aspects, after which the advantages/disadvantages of the approach will be discussed. We end this paper with a section on related work where we compare the aspect-oriented approach with other existing techniques.

2. Security as an aspect

For many problems, the modularity as offered by Object-Oriented programming is not powerful enough to clearly capture some important design decisions, which results in code that is scattered throughout the program. As a short introduction to aspect-oriented programming, we will briefly discuss one particular system, AspectJ [asp, 2001]. This is a Java language extension to support the separate definition of crosscutting concerns. In AspectJ, *pointcuts* define specific points in the dynamic execution of a java program. Pointcut definitions are specified using primitive pointcuts such as the reception of a method call, the creation of a specific type of object, etc. Primitive pointcuts can be combined using logical operators.

On pointcuts, *advice* can then be defined in order to execute certain code. AspectJ supports before and after advice, depending on the time the code is executed. In addition, around advice enables the combination of the former and the latter into one advice. The use of the pointcut and advice constructs will become clearer when we discuss a concrete example.

The definition of pointcuts and the specification of advice on these pointcuts together form an aspect definition. Besides these special constructs [1], an aspect is similar to a class and can as such contain data members, methods, etc. To deploy the aspects in a concrete application, AspectJ provides a special compiler that parses all application and aspect code and transforms them into normal Java code, which is then compiled using a standard Java compiler.

The technique of aspect-oriented programming helps us considerably with the problems described in the introduction. On the one hand, it provides a mechanism to combine separate pieces of code easily, which encourages the separate implementation of non-functional issues like security. Using this divide and conquer strategy, the overall complexity of the problem is reduced considerably. Moreover, it allows different specialists (e.g. an application engineer, a security engineer, ...) to work simultaneously and to concentrate on their field only.

On the other hand, security concerns are often interwoven throughout the application. Aspect-oriented programming is particularly aimed at these crosscutting concerns. It enables interweaving these concerns into the application based on particular rules and automates as such a difficult task that is normally performed by the programmer manually. Hence, it eases the reflection about logically joint, but physically distinct parts.

2.1. An example: Access Control

The example presented in this section discusses how to perform access control in an application. We have chosen this problem because it very clearly shows that security related code can be separated from the functionality of the application in an elegant way.

Basically, access control can be described as follows: at a certain point, the application asks the user to authenticate himself, after which it can allow/deny access based on his identity. However, this abstract view hides several difficult details. The key to turn the above description into an aspect-oriented application is the identification of the important domain concepts and their mutual dependencies.

First, what is the exact entity that has to be authenticated? From a user-oriented view[2], the user of the global application might be a reasonable decision here. In this case the user has to login once, after which this identity is used during the rest of the application. However, the granularity of this approach will clearly not suffice for some applications, e.g. a multi-user or a multi-agent system. A second approach consists of linking the identity to a certain object in the application. Here, login information will be reused as long as the actions are initiated by the same object. However, the identity of the user might change over time. It is then necessary to associate the identity with the initiator of a certain action. In this case, an authentication procedure is required every time the specific action is initiated.

Next, for what resources do we want to enforce access control? Again, one can think of different scenario's. An identity might require access to one resource instance (e.g. a printer). When more instances are available, one could have access to the whole group or to only a particular subgroup. In case of different resource types the identity could require access to a specific combination of these resources. In general, this will often correspond to a combination of (some parts of) application objects.

A last but not less important consideration deals with specifying where the two previous concepts meet each other. This path from the authenticated entity to the resources is necessary to pass login information to the access control mechanism. In a distributed system for instance, authentication and access control might be performed on different hosts. In that case, authentication information must evidently be passed to the access control mechanism in order to ensure correct execution. One obvious example of such access path is the invocation of a specific service of a resource.

Each of the above concepts (identity, resource and access path)is actually a crosscutting entity to the application and maps closely to an aspect. In fact, the three concepts capture the conceptual model of access control and they can as such be used for every access control problem. Note however that we did not discuss any issues concerning concrete mechanisms for authentication and access control. Although certainly relevant, it is important to realize that these are implementation decisions and they will as such depend on the underlying security architecture.

Figure 1 shows the details of one particular case of access control, where each object is authenticated once and where access is checked for each invocation of a particular service. The implementation of the other discussed access control mechanisms would be fairly similar. The observant reader will notice that the aspect code is written for a minimal application that consists of a *Server* implementing a *ServerInterface* with a method *service* and a *Client* invoking this *service*.

The **Identification** aspect is used to tag the entities that must be authenticated. In this case, every object of the class Client is considered as a possible candidate. Furthermore, the aspect implementation contains a field Subject that is used to store the identity information. As such, this information will be available as if it were glued to the particular Client object.

The authenticationCall pointcut of the **Authentication** aspect specifies all places where the service method of the ServerInterface is invoked. Through the use of cflowroot[3] the Authentication aspect travels along with the invocation. Before the method is actually invoked, the identity information from the Identification aspect is copied to a local field of this aspect. As such, it is able to pass the authentication information to the access control mechanism. If the Client was not yet authenticated, this is the right place to do this.

Finally, the **Authorization** aspect checks access based on the identity information received through the Authentication aspect. This check is performed for every execution of the service method (checkedMethods pointcut). In this example, as you can see, the login and access control phase are written in pseudo code. The actual code will depend on the underlying security architecture as discussed before. In our implementation, we have used the Java Authentication and Authorization Service [Lai et al., 1999] for this purpose.

Weaving the above aspects into the application will result in a new, more secure version of the application. In the latter, the access controlling code defined in the Authorization aspect will be executed before every invocation of *service()*. At this point, the application will continue its normal execution if access is granted, however an exception will be

```
aspect Identification of eachobject(instanceof(Client)) {
    public Subject subject null ;
}

aspect Authentication of eachcflowroot(authenticationCall()) {
    private Subject subject;

    pointcut serviceRequest() : calls(ServerInterface, * service(..)) ;

    pointcut authenticationCall() :
                    hasaspect(Identification) && serviceRequest() ;

    before(Object caller) : instanceof(caller) && authenticationCall(){
        final Identification id = Identification.aspectOf(caller);
        if(id.subject == null) {
            <login> ;
            subject = id.subject ;
        }
    }

    public Subject getSubject(){
        return subject ;
    }
}

aspect Authorization {
    pointcut checkedMethods() : executions(* service(..)) ;

    before() returns Object : checkedMethods() {
        Authentication au = Authentication.aspectOf() ;
        Subject subject = au.getSubject() ;
        boolean allowed = <check access control> ;
        if (allowed){
            return proceed();
        }
        else{
            throw new Exception("Access denied") ;
        }
    }
}
```

Figure 1. Aspect code for object-based access control

thrown if the (un)authenticated entity is not allowed to do so. As such, conventional[4] use of the method *service()* will be restricted to certain users depending on the security policy, just as would have been the case by coding the access control mechanism directly into the applition code.

2.2. Generalization of the example

The deployment of each of the crosscutting entities described in the previous section depends heavily on the actual type and implementation of the particular application. For example, an email client will work on behalf of one user, while a multi-user agenda system will want to distinguish his users. Also, objects representing a user will clearly differ in structure and behaviour between separate applications. In general, it is impossible to define one set of aspects that will be applicable to all possible applications. Therefore, a more generic mechanism is desirable that separates the implementation of security mechanisms from these choices.

Given the previous example you might notice that the deployment decisions are actually contained in the pointcut definitions, which define where and when an advice or an aspect has to be applied. For this purpose, AspectJ provides us with the possibility to declare pointcuts abstract and afterwards extend them to define the actual join points. Using this mechanism, it is possible to build a general authorization aspect and redefine the included abstract pointcuts depending on a specific application. To illustrate this technique, we have applied it to the experiment of the previous section. The result is shown in figure 2. In order to use these generic aspects in a concrete situation, one has to extend the abstract aspects and fill in the necessary pointcuts depending on the specific security requirements of the application.

A major advantage of this generalization phase is the ability to reuse the core structure of the security requirement. Since this will be similar for every situation, it is not necessary to reinvent the wheel for every case. It should be properly designed by a qualified person only once, after which aspect inheritance enables easy reuse.

2.3. Towards a framework of security aspects

For a secure distributed application, other security requirements besides authentication and authorization must be considered, such as confidentiality, non-repudiation etc. We will now briefly describe how they could be implemented using aspects.

Encryption of objects is required for *confidentiality* and *integrity*. This is a quite straightforward task using the Java JCA/JCE [Gong, 1998]. Two issues have to be considered. First, one has to decide where and how to insert this into the application. One possibility is to encrypt objects while they are written to a specific stream. For this case, the stream can be wrapped by a specific encryption stream. Another possibility is to encrypt objects whenever they are serialized. Therefore, the readOb-

```
abstract aspect Identification of eachobject(entities()) {
    abstract pointcut entities() ;

    public Subject subject null ;
}

abstract aspect Authentication of eachcflowroot(authenticationCall()) {
    private Subject subject;

    abstract pointcut serviceRequest() ;

    ...
}

abstract aspect Authorization {
    abstract pointcut checkedMethods() ;

    ...
}
```

Figure 2. Generalized aspect code for access control

ject() and writeObject() methods of the object should be overridden to include encryption here. Second, there is the issue of how to get or store the cryptographic keys. Similar to the identity in the previous section, one has to find some entity in the application with which the keys will be associated. The implementation will vary according to how the keys are to be acquired.

Non-repudiation requires the generation of proof for certain events in the system, e.g. the invocation of a specific method. This is quite similar to the problem of access control described above. One crosscutting entity defines the identity that wants to generate the proof. Another entity stores and manages these proofs. And finally, a third entity defines where and how proofs should be generated and passed along.

In the end, a combination of all the security aspects could form the basis of an aspect framework for application security. This framework will consist of generalized aspects for each of the security requirements. Note that several aspect implementations, depending on different underlying security mechanisms, may be included for the same security requirement. The deployment of the framework for a concrete application will then come down to choosing the necessary aspects and defining concrete pointcut designators for them. This technique actually suggests three distinct tasks to develop a secure application [5] : build the application, develop a generic security aspect architecture and specify the aspect deployment pointcuts. A more elaborate discussion on this security framework can be found in [Vanhaute et al., 2001].

3. Discussion

Being able to specify security concerns in a separate way, and still having them applied throughout the whole application is a noble goal. Aspect oriented programming techniques hold a promise of achieving this goal.

Although the technology has not yet fully matured, the current possibilities of AspectJ already give us a number of interesting advantages:

- Security should be applied at all times, if it is to be applied correctly. By looking at the definition of the pointcuts in the aspect that implements that particular security concern, a security engineer immediately knows all the places where this concern will be used, given that the AspectJ compiler does its job correctly.
- The implementation of the security mechanisms does not have to be copied several times. All the implementation code can be gathered within a small number of advices, perhaps all within one source file. As a result, when changes have to be made, the programmer can focus on that one part.
- Another result of the separate specification of the aspect code is that management of the different packages of an application is easier. There can be separate packages for pure application functionality, one for pure security code and a package that defines the points where the security code is to be applied.
- Although it could be argued that simple text substitution tools would also be able to insert code in a generic way into a program, the aspect oriented approach has much less chance of introducing bugs. The constructs aspect oriented transformers work on, are not mere text elements, but language constructs. These map more naturally onto the entities a security policy would speak about.
- By making the aspects more generic with respect to an application, we obtain a good combination of both application independent implementations of security concerns and the use of these implementations within the context of a specific application. The support for abstract pointcuts within AspectJ makes it possible to specify the two in separate files.

For our work, the use of the current version of AspectJ (0.7) has also some drawbacks (see further). On the one hand there are some technical issues regarding the current implementation of AspectJ. It is expected that these will be solved in later releases of the tool. On the other hand there are problems that are more fundamental in nature. For this, at least a redesign of the aspect-oriented tool is needed.

- For each method call that has some security concern, the AspectJ compiler will insert one or more extra calls. Therefore, the generated code is less efficient, and introduces more overhead than a direct implementation would have. Unfortunately, this is the price to pay for the genericity of our approach. However, it is certainly not worse than some other systems discussed in the next section. Building a more complex, but less general aspect combination tool could solve this.

- If not all code in the application can be trusted, one has to be very certain the generated code does not add any security holes. For instance in the case of authorization: it should not be possible for a client to call the end-functionality of a server through some other, by the aspect tool generated method, in order to circumvent the authorization checks. This means the security implementer has to have a very clear idea of what and how exactly the aspect tool produces. At this moment, the output of the AspectJ compiler cannot be trusted yet, because the original functionality, without the new aspect code, is only moved into a new method with a special name. However, this is only a problem if not all source code is under the control of the AspectJ compiler. The fact that AspectJ is not a formal proven language only increases this problem. This might become one of our topics for future work.

- Another issue related to the generated code comes up when debugging. As the runtime code does not directly correspond to the code the programmer wrote, it can be harder to figure out what is going on. The aspect-oriented research community is at the moment trying to build better support tools that would help the programmer in relating runtime events with the source code it stems from.

- The implementer of the security code still has to have very detailed knowledge of security mechanisms, their strong and weak points, how to implement them. As AspectJ is a generic tool, it does not help the programmer here, apart from providing a better modularization of the problem. However, this is not a particular problem of AspectJ, but rather of our approach to the problem.

4. Related work

There are already a large number of security architectures proposed or implemented in Java, e.g. [De Win et al., 2000]. Sun for one has for instance released JAAS [Lai et al., 1999] for authentication and authorization, SSE for secured network communication, and there are proposals for a secure RMI implementation. These will often already realize the intended result, and can therefore be used in the implementation of the security aspects. The combination of existing technologies with aspect-oriented programming is not expected to pose severe problems. The added value of aspects in this case is the possibility to have a much more flexible security policy, and this at a granularity that corresponds better with the application, i.e. at the level of method calls and objects. Some of the proposed architectures also have a fine granularity, but the configuration and mapping onto what happens inside an application can be fairly difficult.

By using a number of object-oriented design patterns [Gamma et al., 1994], the existing security architectures also try to be independent of an application structure, and they all succeed in this to some degree. The drawback of this design is that the structure of the solution becomes more complex and harder to understand. With an aspect-oriented approach these implementations can be designed in a more natural way.

Transformations in AspectJ happen on the level of source code. Other tools are available that work on the level of byte code [Cohen et al., 1998, Keller and Holzle, 1998]. This has the advantage that you can add your own aspects even when no source code is available for the application. The disadvantage is that on the level of byte code, a lot of the application logic is already lost. Reconstructing this is often hard, and giving correct descriptions of how a series of byte codes has to be changed to for instance implement authentication will be even harder. Checking and debugging the result will also be difficult.

There is also research into a more declarative description of security properties for an application [Evans and Twyman, 1999, Hagimont and Ismail, 1997]. This corresponds to an aspect-oriented language of the first type of section 2. The real challenge here is to think of the right abstractions the description will consist of. This is not at all an evident matter, certainly if a goal is to be generic. We think it is better to first experiment with a generic aspect-oriented language as described in this paper. From these experiments, we would hope to distill the important abstractions.

Meta level architectures [Chiba, 1995, Robben et al., 1999, Stroud and Wue, 1996] also make it possible to separate application from security implementation [Ancona et al., 1999, Welch and Stroud, 2000]. They offer a complete reification of what is going on in the application: the events of sending a message, starting the execution, creating an object all get reified into a first class object. Because the meta-program has control over these reified entities, it can intervene in the execution of the base application. In comparison to aspect-oriented programming this mechanism is much more powerful, but it is also heavier. Moreover, the development of meta-programs for security is more complex, because the programmer is forced to think in terms of meta-elements, which is only indirectly related to the application.

Other approaches [Fraser et al., 1999] also use the basic idea of introducing an interceptor between clients and services, for instance to do extra access control checks. They are similar to meta-level architectures in that they also intervene in the communication between client and service, but the intervention is less generic (and heavy): the interceptors are mere decorators around the services. In simple situations, they can be specified fairly easy, perhaps through some declarative description. However, when more and more application state need to be taken into account, writing decorators becomes very hard, or even impossible due to the bounded possibilities of the declarative language.

5. Summary

This paper presented the use of aspect-oriented programming to add security to an application. By means of the example of access control, we first demonstrated the feasibility of this approach. In order to construct a more generic solution, we suggested to abstract relevant pointcuts out of the aspect implementation. This enabled us to separate the security mechanisms from the actual policy, which promotes the reuse of the mechanism implementations. After briefly discussing some other security requirements, we touched upon the feasibility to build a security aspect framework. Finally, we discussed the advantages and disadvantages of our approach.

The most important advantage of this approach is the separation of the application and the security related code. This considerably simplifies the job of the application programmer. Moreover, the security policies are gathered in one place, which makes it easier to check whether all the requirements are met. Still, we think that the deployment of these generalized aspects remains quite difficult. We would like to focus our research in the future on this issue, for example by automating the generation of concrete pointcuts based on a simplified high level description.

Notes

1. AspectJ also supports other constructs like Introduction. Since they are not used directly in this paper, we will not discuss them here.

2. From another point of view, application code might be the subject of authentication. While the mechanism to establish the correct identity of the code originator might be different, the overall authorization mechanism described in this paper will still be applicable.

3. cflowroot is a predefined keyword in AspectJ that denotes every control flow leading to that particular pointcut. Using this keyword, it is possible to 'follow' the invocation stack and pass as such information from the caller to the callee.

4. By predicting the output of the aspect weaver, one might be able to circumvent this access control mechanism under certain circumstances. We discuss this problem in detail in section 3.

5. A similar separation of these tasks has been described in [Robben et al., 1999].

References

[asp, 2001] (2001). AspectJ Website. http://www.aspectj.org/.

[cer, 2001] (2001). CERT Website. http://www.cert.org/.

[Ancona et al., 1999] Ancona, M., Cazzola, W., and Fernandez, E. (1999). Reflective Authorization Systems: Possibilities, Benifits and Drawbacks. In *Secure Internet Programming: Security Issues for Mobile and Distributed Objects.*

[Chiba, 1995] Chiba, S. (1995). A MetaObject Protocol for C++. In *Proceedings of the 1995 Conference on Object-Oriented Programming.*

[Cohen et al., 1998] Cohen, S., Chase, J., and Kaminsky, D. (1998). Automatic Program Transformation with JOIE. In *Proceedings of the 1998 USENIX Annual Technical Symposium.*

[De Win et al., 2000] De Win, B., Van den Bergh, J., Matthijs, F., De Decker, B., and Joosen, W. (2000). A Security Architecture for Electronic Commerce Applications. In *Information Security for Global Information Infrastructures*, pages 491–500. IFIP TC11, Kluwer Academic Publishers.

[Demailly, 1996] Demailly, L. (1996). Netscape Security (problems). http://www.demailly.com/ dl/netscapesec/.

[Evans and Twyman, 1999] Evans, D. and Twyman, A. (1999). Flexible Policy-Directed Code Safety. In *Proceedings of the 1999 IEEE Symposium on Security and Privacy.*

[Fraser et al., 1999] Fraser, T., Badger, L., and Feldman, M. (1999). Hardening COTS Software with Generic Software Wrappers. In *Proceedings of the 1999 IEEE Symposium on Security and Privacy.*

[Gamma et al., 1994] Gamma, E., Helm, R., Johnson, R., and Vlissides, J. (1994). *Design Patterns: Elements of Reusable Object-Oriented Software.* Addison Wesley Longman.

[Gong, 1998] Gong, L. (1998). Java Security Architecture. http://java.sun.com/security.

[Hagimont and Ismail, 1997] Hagimont, D. and Ismail, L. (1997). A Protection Scheme for Mobile Agents on Java. In *Proceedings of the International Conference on Mobile Computing and Networking.*

[Keller and Holzle, 1998] Keller, R. and Holzle, U. (1998). Binary Code Adaptation. In *Proceeding of the 1998 European Conference on Object-Oriented Programming.*

[Lai et al., 1999] Lai, C., Gong, L., Nadalin, A., and Schemers, R. (1999). User Authentication and Authorization in the Java Platform. In *Proceedings of the 15th Annual Computer Security Applications Conference.*

[Robben et al., 1999] Robben, B., Vanhaute, B., Joosen, W., and Verbaeten, P. (1999). Non-functional Policies. In Cointe, P., editor, *Meta-Level Architectures and Reflection*, volume 1616 of *Lecture Notes in Computer Science*, pages 74–92. Springer-Verlag.

[Stroud and Wue, 1996] Stroud, R. and Wue, Z. (1996). Using Metaobject Protocols to Satifsy Non-functional Requirements. In *Advances in Object-Oriented Metalevel Architectures and Reflection.*

[Vanhaute et al., 2001] Vanhaute, B., De Win, B., and De Decker, B. (2001). Building Frameworks in AspectJ. ECOOP2001 Workshop on Advanced Separation of Concerns.

[Welch and Stroud, 2000] Welch, I. and Stroud, R. (2000). Using Reflection as a Mechanism for Enforcing Security Policies in Mobile Code. In *Proceedings of the Sixth European Symposium on Research in Computer Security.*

EXTENDING A CAMPUS NETWORK WITH REMOTE BUBBLES USING IPSEC

Aurélien Bonnet
Université catholique de Louvain, Dept INGI
Place Ste-Barbe, 2, 1348 Louvain-la-Neuve, Belgium
ab@info.ucl.ac.be

Marc Lobelle
Université catholique de Louvain, Dept INGI
Place Ste-Barbe, 2, 1348 Louvain-la-Neuve, Belgium
ml@info.ucl.ac.be

Abstract

There is an increasing demand for high speed remote connections (e.g. ADSL, cable, etc.) to university networks both from students and staff members. In many instances, the need is to connect not a single computer but a remote subnetwork and to be able to have not only clients but also servers on the remote subnetwork (for instance to access files on computers on the remote network from workstations in the university). We present here a solution to meet these needs through a simple dial-up-like connection to which the access provider allocates only a single temporary IP address to a single remote machine. The solution allows remote networks to be started anytime like little bubbles and be integrated dynamically into the big bubble that is the university network. It is based on the use of IPSec, DHCP and address translation. Beside providing confidentiality, this allows allocating dynamically IP addresses in the range of the university to computers in the remote bubble, and binding them to permanent DNS names in the domain of the university. After configuration, the computers in the remote subnetwork appear as if they were located inside the university. The solution can, of course, be applied to any organization.

Keywords: Remote networks, Virtual private networks, IPSec, DHCP, NAT, RSIP.

Introduction: Remote access to university networks

New learning technologies intensively use computers and computer networks. Universities have to provide an increasing number of workstations and personal computers to their students. All these computers are connected to the university network, and through this, to the Internet.

On the other hand, an increasing number of students own or have access to a computer at home or in their residence near the university. In many instances, more than one computer is available: when a new computer is bought, the old one is rarely discarded, but it is rather restricted to the less demanding applications (E-mail, web access, word processing), so that several members of a household can be simultaneously working on computers. In multi-room flats used as student accomodation, several computers are also often available.

However, many learning tasks cannot be performed at home because they require permanent or quasi permanent access to servers of the university, or simply to the Internet. Other activities are only possible with computers that are part of the university network, because of, for instance, software licensing restrictions.

The availability of high speed remote connections (ADSL, Cable) makes it possible to have at home the same kind of access speed as in one's office or in a room of workstations. However, all the other restrictions of dial-up connections remain since remote users get only one dynamically allocated IP address in the range of the provider.

In this paper, we shall discuss how to connect a remote subnetwork when the provider allocates only one dynamic IP address to it. The purpose is to make the remote subnetwork, called remote bubble, really look like part of the university network, with IP addresses belonging to this network. These addresses are centrally and dynamically allocated (central and dynamic allocation avoids wasting IP addresses). As far as we know, this problem has never been addressed before.

1. Connection of a subnetwork through a single IP address: state of the art

There are different solutions to connect subnetworks through a single IP address. The first is to use NAT/PAT (Network Address Translation and Port Address Translation); software as well or hardware products (soho routers) implementing this solution are available. They are often based on Linux IP masquerading. Other solutions use tunneling protocols to build a Virtual Private Network.

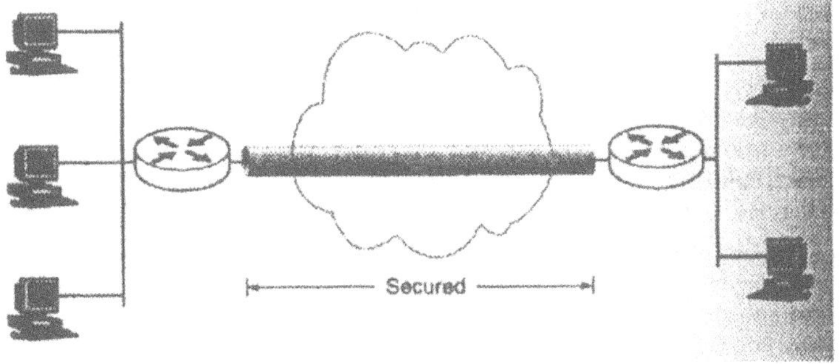

Figure 1. A VPN across the Internet

1.1. IP Masquerading

IP Masquerade is a Linux networking function similar to one-to-many NAT (Network Address Translation) found in many firewalls and network routers. For example, if a Linux host, called the gateway, is connected to the Internet, the IP Masquerade feature allows "client" applications on other computers connected to the internal interface of this gateway and using non-routable IP addresses to reach the Internet.

This system allows a set of machines to invisibly access the Internet through the gateway. To other machines on the Internet, all this traffic will appear to be from or to the gateway (for more information see [Ran00]). Not even all "client" applications work with this scheme (e.g. FTP). It is therefore often complemented with specified application level gateway programs.

1.2. Building Virtual Private Networks

The problem that VPNs are trying to solve is that of letting two networks communicate securely when the only connection between them is over a third network which they don't trust. VPNs use a gateway between each of the communicating networks and the untrusted network. Most of the current VPN packages use tunneling to create a private network. The principle of tunneling is to encapsulate a packet within a packet.

The gateway machines can encrypt packets entering the untrusted net and decrypt packets leaving it, creating a secure tunnel through it (see figure 1).

1.2.1 Simple tunneling protocol. Most of the current operating system can enable simple tunnels between two gateways (without any authentication or encryption: the tunnel is thus not secure). This system is very simple. Gateways on each network encapsulate packets destined to the distant network in a packet destined to the remote gateway. But this solution is limited to static IP addresses on gateways. In our context, IP addresses of gateways will be dynamically allocated by the provider. So we need to authenticate these gateways dynamically and by other mean than its IP address.

1.2.2 The IPSec Protocol and its use in Virtual Private Networks . IPSec is a mechanism for adding security to IP and upperlayer protocols. It can protect traffic between hosts, between network security gateways (routers, firewalls,..) and between hosts and security gateways. IPSec hosts and gateways are authenticated by cryptographic technics independantly of their IP addresses, which can be allocated dynamically.

IPSec defines two different protocols: The Encapsulating Security Payload (ESP), and the Authentication Header (AH). AH provides proof-of-data-origin, data integrity and antireplay protection on received packets. ESP provides, in addition, data confidentiality and limited traffic flow confidentiality.

More information on the IPSec protocol, can be found in [DH99], [RFC2401], which defines the base architecture upon which all implementation are built, [RFC2402] which explain AH functionalities, and [RFC2406] which talks about ESP.

The VPN can be built by deploying IPSec gateways. The protected network to which access is controlled is on one side of the gateway; the unsafe and unsecured network (usually Internet) is on the other. IPSec must be used in tunnel mode between the gateways, because the VPN is protecting the traffic between two different networks.

2. Using IPSec VPNs to connect remote bubbles to a university network

We want to integrate computers at home (or in student residences) transparently in the university network. The homes or the residences are connected to a provider and obtain from it a single temporary IP address: this IP address may be different each time they connect. Several uses of this single temporary IP address are possible:

- One single computer is connected to the modem (ASDL/Cable), and uses the temporary IP address. This solution is the easiest,

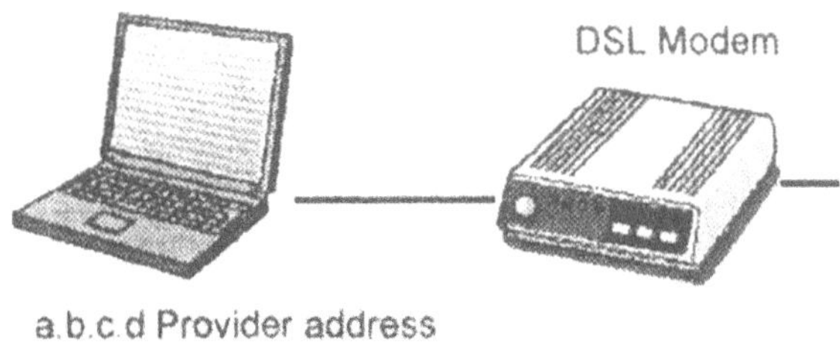

Figure 2. Single connection with a DSL Modem

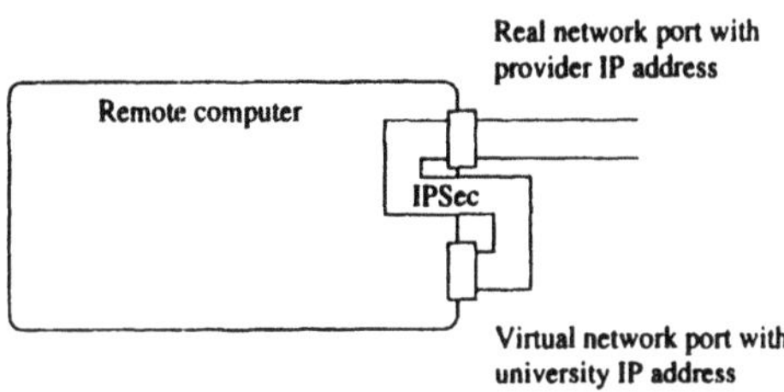

Figure 3. Using IPSec to add an address of the university network to a remote computer

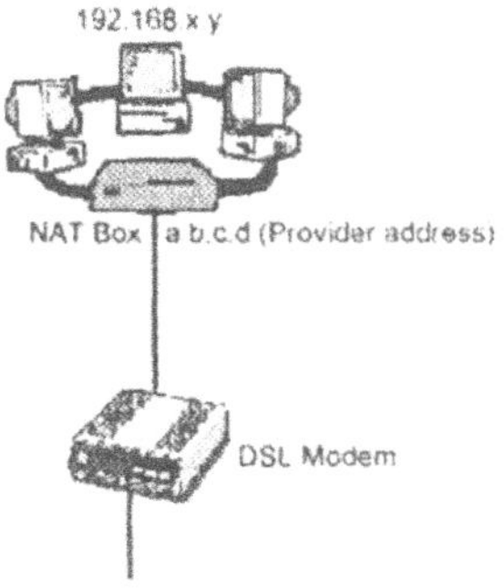

Figure 4. Architecture of a NAT network

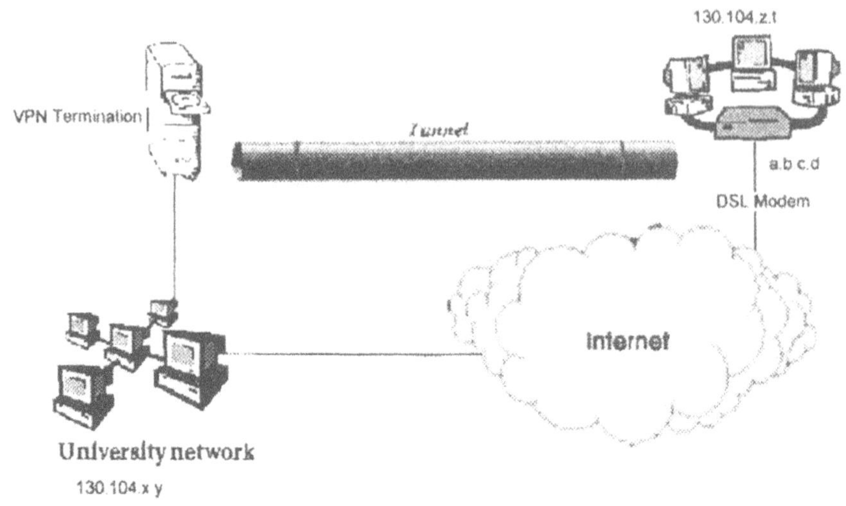

Figure 5. Remote bubble with IPSec

but the computer appears with a provider address, not an address of the university network, as requested (see figure 2). To overcome this problem, an internal address of the university network could be allocated to a virtual interface of the computer and routed through an IPSec VPN across the provider network to a gateway in the university (See figure 3). This is actually a particular case of the third situation.

- One computer is a gateway, uses IP masquerading, and connects a subnetwork to the Internet (a dedicated NAT box is equivalent to this. See figure 4). This solution allows to connect more than one computer through a connection with a single routable IP address. It is integrated in Linux, free, and easy to build (several companies market black-boxes with this kind of functionality), but the machines in the subnetwork are not accessible from the Internet (they can only be clients, not servers), and some protocols are not compatible with it (H323, IPSec,...).

- One virtual private network is set up between a gateway to the remote subnetwork and a gateway to the university. This solution doesn't impose any constraint on the address allocation policy in the remote subnetwork: the computers of the remote subnetwork can have routable IP addresses. This solution is the only one allowing "remote bubbles" to merge transparently in the university network (see figure 5). That's why we will develop this solution in this section.

2.1. Requirements of the VPN

The VPN used to connect the remote subnetwork to the university network must meet the following requirements:

- We want to be able to decide from within the university what addresses will be allocated to the computers on the subnetwork. These addresses may be non routable or routable (for instance taken from the university range).
- We want also to be able to allocate these addresses dynamically but to be bound to permanent domain names. We will not discuss this last aspect in this paper.
- We want the remote network to merge automatically in the university network when the gateway is started. The computers on the remote network must then be undiscernible by third parties from computers located inside the university.
- We want the same security beween computers in the remote network and other computers in the same department of the university as between computers in this department.

2.2. SubNetwork with IPSec Gateways

Using an IPSec tunnel between the gateway to the remote subnetwork (let's call it "*Hawser*") and a gateway in a department of the university (let's call it "*Bollard*") already allows to meet all but the second of the above requirements (see figure 6).

- The computers of the subnetwork are logically neighbours of the other end of the tunnel.
- IPSec is a standard protocol available for any decent platform.
- IPSec authentication of the gateway is not based on its IP address that can be dynamically allocated by the provider.
- IPSec provides the required security. If confidentiality is needed, ESP can be used, otherwise AH is sufficient. If security is definitely not a problem (for instance if the ADSL links are connected

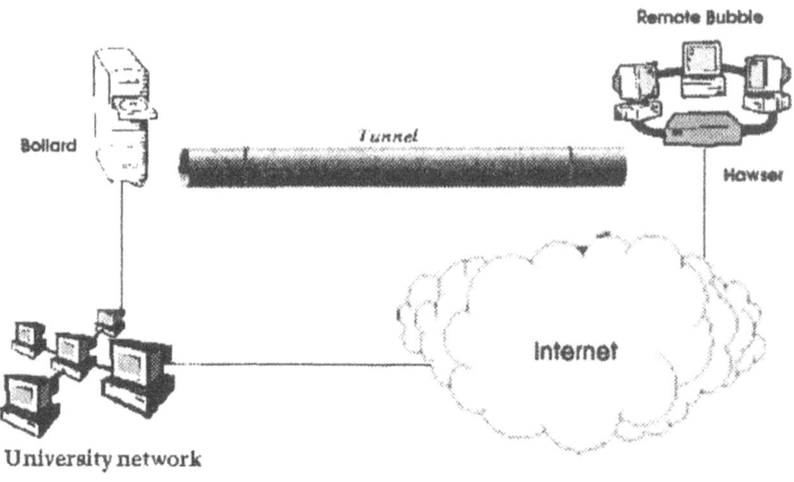

Figure 6. Architecture of the solution

directly by the provider to the university through a VPN instead through the public Internet), then AH with null encoding can be used. The choice is a matter of balance between security and performance. Benchmarks will tell us if providing less than maximum security makes any difference from a performance point of view.

Meeting all these requirements is not straightforward. Let us look more closely at what should happen when *Hawser* boots:

- *Hawser* boots.
- *Hawser* logs into the provider network and gets a dynamic provider address from, say, a radius server.
- *Hawser* calls *Bollard* and sets up an IPSec tunnel.

All the packets destined to the subnetwork will be routed through *Bollard* and *Hawser*, and all external packets from the subnetwork will be routed through *Hawser* and *Bollard*.

3. Dynamic allocation and address translation

The Dynamic Host Configuration Protocol *(DHCP)* automates the process of configuring devices on IP networks. DHCP performs many of the functions a network administrator could carry out manually when connecting a new computer to a network (see [DL99]). With DHCP relay agents, remote machines can also be configured. We decided to use the DHCP protocol with relay agents to configure the differents subnetworks for differents reasons:

- addresses can be leased temporarily when needed which simplifies network administration of nomadic computers (laptops),
- subnetworks can be created without any administrative overhead for address allocation,
- the network configuration of the computer is easier (most of the parameters are transmitted by the protocol),
- the DHCP protocol is available on many operating systems.

3.1. The relay agent

A relay agent is designed to forward DHCP messages between clients and server when the server and the client are not in the same network. When the relay agent receives a client message, it forwards it to the server, and when the server answers to a request, it also uses the relay agent to contact the client. In our solution, the relay agent runs on *Hawser*, and the DHCP server can run on any machine of the same network as *Bollard*, but it is easier to put it directly on *Bollard*.

3.2. DHCP configuration

As specified in [RFC2131], the technique is to assign statically a range of addresses (subnetwork class) to each network behind a relay agent. This solution would be the easiest to deploy but it wastes a lot of IP addresses.

With a modified relay agent, another solution would be to assign addresses to the devices on the different subnetworks without regard for their localization. This solution is more economical in IP addresses, but it is more difficult to deploy, since routes must be explicitly configured for each individual device.

When a device asks for a new DHCP configuration, a dedicated IPSec tunnel must be opened between *Bollard* and *Hawser* for this new IP address. This will be done by the relay agent, as it has all the required informations.

3.3. Address translation

The solution proposed in the previous paragraph has a problem. All the packets sent by a device on a subnetwork will be routed through *Bollard* and *Hawser*, even the packets destined to same remote bubble, as the differents devices are not considered to be on the same virtual subnetwork. The solution we use to avoid this problem is the address translation (More information available on [Dut01]).

Instead of sending address like "a.b.c.d", the DHCP server will send address like "10.b.c.d" with a class A subnetwork mask. Thus all devices in the same remote bubble can see each other. *Hawser* will have to "NAT" (translate address) 10.b.c.d and a.b.c.d for incoming and outgoing messages. Devices in other bubbles are known with their addresses allocated and not "10.b.c.d".

3.4. Summary

When a new computer is started on a remote bubble:

- the computer boots.
- it broadcasts a DHCP request.
- *Hawser* forwards the request to *Bollard* (assuming it is the DHCP server).
- *Bollard* replies to the request with the address 10.b.c.d
- *Hawser* forwards the reply to the computer, starts address translation between 10.b.c.d and a.b.c.d, and enables the tunnel for the address a.b.c.d
- the computer is connected to the network.

When a computer uses the network:

- if the address is local (10.x.x.x) it connects directly the specified computer, else it sends the packet to *Hawser*, which it considers as its default router.
- *Hawser* exchanges the source address (10.b.c.d to a.b.c.d)
- *Hawser* sends the new packet to *Bollard* through the tunnel.
- *Bollard* routes the packet it receives like a normal packet.

When a computer wants to connect or reply to a computer on a remote bubble:

- the packet is sent to *Bollard.*

- *Bollard* sends it to the *Hawser* concerned (several remote bubbles can be attached to the same *Bollard* via different *Hawser*) through a tunnel.

- *Hawser* translates the destination address (a.b.c.d to 10.b.c.d).

- *Hawser* forwards the packet to the computer concerned in the remote bubble.

4. Related Works

We did not find any other published paper on the problem of setting up a remote subnetwork connected with a single temporary address allocated by a provider and using addresses belonging to a main network and dynamically allocated by a server in this network. However, results has been published about closely related problems.

Realm Specific IP (RSIP) is a new protocol, designed as an alternative to NAT that preserves end to end packet integrity. RSIP allocates on demand in a network A, addresses belonging to another network B. The gateway between networks A and B has a pool of IP addresses of B and will allocate them to hosts in A (called RSIP hosts) when requested. Routes or tunnels must be provided in A for these RSIP hosts. They can be considered as if they were really on the network B.

The advantage of this solution over NAT is that nothing in the packet is modified. The drawback is that a driver is needed in each RSIP host of the subnetwork.

RSIP was designed to solve a different problem than the one addressed in this paper. However, it provides a similar service with different mean. The main diference of the service provided by RSIP and the service provided in our solution is that, since the RSIP server is on the gateway, the range of addresses allocated to each remote subnetwork must be allocated statically. Since our solution uses a central server, addresses can be distributed on demand to several remote networks, which is a more efficient allocation technique.

On the other hand, RSIP preserves the end to end integrity of packets while our solution uses NAT coupled with DHCP which means that packets are modified in the path. The price RSIP has to pay for this advantage is to add some software to each host in the subnetwork.

More Information on RSIP can be found in [RSIP-PROTO].

5. Conclusion

The paper identifies a new problem, which is to make a remote subnetwork, called remote bubble, really look like merged in another network, with IP addresses belonging to this network centrally and dynamically allocated.

It proposes a solution based on existing protocols (IPSec, DHCP, NAT). Coupled DHCP relay and NAT protocol engines provide the hosts of the remote bubble with addresses from the main network. IPSec allows the gateway itself to have a dynamic IP address and hide the contents of packets on the way between the remote bubble and the main network. The situation of the hosts in the remote bubble is exactly the same (except for communications delays) as if they were located inside the main network.

The solution we propose presents many advantages. In addition to secure communication, with the creation of an IPSec tunnel between the remote bubble and the university network, its dynamic configuration allows to manage easily many bubbles of any size, without wasting IP addresses.

The administration of a bubble is quite simple: all the computers plugged behind the gateway (*Hawser*) will be automatically on the university network.

Acknowledgements

This work has been performed in the framework of the SIN project (Secure IP Networking) supported by grant FIRST 981/3749 by MRW-DGTRE ("Travaux réalisés avec l'aide du Ministère de la Région Wallonne, en application du programme de formation et d'impulsion à la recherche scientifique et technologique").

The authors express their thanks to the anonymous reviewers for their useful comments.

References

[Ran00] Ranch, D. (2000). *Linux IP Masquerading HOWTO.* Technical report. http://www.linuxdoc.org/HOWTO/IP-Masquerade-HOWTO.html.

[DH99] Doraswamy, N.; Harkins, D. (1999). *IPSec : The New Security Standard for the Internet, Intranets and Virtual Private Network.* Prentice Hall PTR.

[RFC2401] Kent, S.; Atkinson, R. (1998). *Security Architecture for the Internet Protocol.* Request for Comments 2401. Network Working Group.

[RFC2402] Kent, S.; Atkinson, R. (1998). *IP Authentication Header (AH).* Request for Comments 2402. Network Working Group.

[RFC2406] Kent, S.; Atkinson, R. (1998). *IP Encapsulating Security Payload (ESP).* Request for Comments 2406. Network Working Group.

[RFC2131] Droms, R. (1997). *Dynamic Host Configuration Protocol (DHCP).* Request for Comments 2131. Network Working Group.

[DL99] Droms, R.; Lemon, T. (1999). *The DHCP Handbook.* Macmillan Technical Publishing.

[Dut01] Dutcher, B. (2001). *The NAT Handbook.* Wiley Computer Publishing.

[RSIP-PROTO] Borella M.; Grabelsky D.; Taniguchi K. (2000). *Realm Specific IP: Protocol Specification.* Draft. Network Working Group.

COMBINING WORLD WIDE WEB AND WIRELESS SECURITY

Joris Claessens, Bart Preneel, Joos Vandewalle
COmputer Security and Industrial Cryptography (COSIC)
Dept. of Electrical Engineering - ESAT
Katholieke Universiteit Leuven
http://www.esat.kuleuven.ac.be/cosic/
joris.claessens@esat.kuleuven.ac.be

Abstract In current electronic commerce systems, customers have an on-line interaction with merchants via a browser on their personal computer. Also payment is done electronically via the Internet, mostly with a credit card. In parallel to this, e-services via wireless-only systems are emerging. This paper identifies security and functionality weaknesses in both of these current approaches. The paper discusses why and how general-purpose mobile devices could be used as an extension to PC based systems, to provide more security and functionality. General-purpose mobile devices are shown to be an alternative to costly special-purpose hardware. This combined approach has in many cases more interesting properties than when using mobile devices only. As an example of the combined approach, a GSM based electronic payment system is proposed and investigated. The system enables users to order goods through the World Wide Web and pay by using their mobile phone.

Keywords: WWW security, wireless security, m-commerce

1. Introduction

In current electronic commerce systems, customers have an on-line interaction with merchants via a browser on their personal computer. Also payment is done electronically via the Internet, mostly by sending a credit card number to the merchant. This basic system is in widespread use today, and most people are familiar with buying books and music, booking flights, ordering PCs, etc. There are however some important security problems. For example, credit card numbers are often stolen by hackers from merchants' computers, orders and confirmations are usually not digitally signed and can be repudiated afterwards. In parallel to the fixed PC based systems, e-services are also emerging in the wireless world. Current mobile devices have however rather limited functionality, and in many applications, they are not suited to be used on their own.

This paper suggests a combined approach in which mobile devices are used as an extension to the World Wide Web environment. The paper starts with a description of the security properties of the World Wide Web in Sect. 2, and the security features in some wireless systems, i.e., GSM and WAP, in Sect. 3. Section 4 discusses security and functionality weaknesses in both worlds, and suggests a combined approach. An example of this approach is given in Sect. 5: a GSM based electronic payment system for the WWW is proposed and investigated. Further analysis of this system is presented in Sect. 6.

2. World Wide Web security

There are many security issues related to the WWW. Within the scope of this paper, we will only discuss the communications security aspect, both at the network and the application level, and the payment security aspect.

2.1. Communications security

The communication between a web browser and a web server is secured by the SSL/TLS protocol. Historically, Secure Sockets Layer (SSL) was an initiative of Netscape Communications. SSL 2.0 contains a number of security flaws which are solved in SSL 3.0. SSL 3.0 was adopted by the IETF Transport Layer Security (TLS) working group, which made some small improvements and published the TLS 1.0 [8] standard. "SSL/TLS" is used in this paper, as "SSL" is an acronym everyone is quite familiar with; however, the use of TLS in applications is certainly preferred to the use of the SSL protocols.

Within the protocol stack, SSL/TLS is situated underneath the application layer. It can in principle be used to secure the communication of any application, and not only between a web browser and server. SSL/TLS provides entity authentication, data authentication, and data confidentiality. In short, SSL/TLS works as follows: public-key cryptography is used to authenticate the participating entities, and to establish cryptographic keys; symmetric key cryptography is used for encrypting the communication and adding Message Authentication Codes (MACs), to provide data confidentiality and data authentication respectively. Thus, SSL/TLS depends on a Public Key Infrastructure. Participating entities should have a public/private key pair and a certificate. Root certificates (the certification authorities' certificates that are needed to verify the entities' certificates) should be securely distributed in advance (e.g., they are shipped with the browsers). Private keys should be properly protected. Note that these two elements, i.e., distribution of root certificates in browsers and the protection of private keys, is actually one of the weak and exploited points with respect to WWW security (see further).

More detailed information on SSL/TLS, the security flaws in SSL 2.0, and the differences between SSL 3.0 and TLS 1.0, can be found in [27].

2.2. Application security

SSL/TLS only protects data while it is in transit. Moreover, exchanged messages are not digitally signed. Therefore it does not provide non-repudiation. Both customers and merchants can always deny later on having sent or received requests or confirmations from each other.

In addition to SSL/TLS, critical messages should thus be digitally signed before they are sent through the secure channel. The concept of digitally signing messages is not really integrated yet in today's web browsers. Netscape though allows the content of forms to be digitally signed using the Javascript `signText()` function. XML will be more and more used on the WWW to represent content instead of the basic HTML. In the future, browsers are therefore expected to implement Signed XML [10], which specifies how XML documents should be digitally signed.

Note that an alternative protocol to secure the communication on the WWW has been proposed in the past: S-HTTP [26]. This protocol is situated at the application layer, and is specifically intended for HTTP. It secures HTTP messages in a very similar way to the protocols for secure email, and provides non-repudiation. SSL/TLS has however become the de-facto standard on the web, and S-HTTP was not a success.

2.3. Payment security

Although numerous different electronic payment systems have been proposed that can be or are used on the WWW, including micro-payment systems and cash-like systems, most transactions on the web are paid using credit cards. Mostly, customers just have to send their credit card number to the merchant's web server. This is normally done 'securely' over SSL/TLS, but some serious problems can still be identified. Users have to disclose their credit card number to each merchant. This is quite contradictory to the fact that the credit card number is actually the secret on which the whole payment system is based (note that there is no electronic equivalent of the additional security mechanisms present in real world credit card transactions, such as face-to-face interaction, physical cards and handwritten signatures). Even if the merchant is trusted and honest this is risky, as one can obtain huge lists of credit card numbers by hacking into (trustworthy, but less protected) merchants' web servers. Moreover, it is possible to generate fake but valid credit card numbers, which is of great concern for the on-line merchants. Thus, merchants bear risk in card-not-present transactions.

Secure Electronic Transaction, SET [29], is a more advanced standard for credit card based payments. One of its core features is that merchants only see encrypted credit card numbers, which can only be decrypted by the issuers. This system is conceptually much better, but until now it has not become popular due to its complexity.

American Express offers a 'one-time credit card' solution [1] with which customers can protect their privacy, but which also solves some of the above mentioned problems. Alternatively, several similar systems exist (e.g., InternetCash [16]) in which customers can obtain some prepaid value identified and protected with a number and PIN, and use it on-line in cooperation with a central server. Finally, real-life electronic payment means (e.g., Proton [25] and debit cards) are also starting to be deployed on the WWW (e.g., [2]).

3. Wireless security

GSM and WAP are currently probably the two most popular and widely used wireless technologies. They are briefly presented in the following paragraphs. Thereafter, some other systems and initiatives in the wireless world are discussed.

3.1. GSM

GSM, Global System for Mobile communications, is the currently very popular digital cellular telecommunications system specified by the European Telecommunications Standards Institute (ETSI). In short, GSM intends to provide three security services [32]: temporary identities, for the confidentiality of the user identity; entity authentication, that is, to verify the identity of the user; and encryption, for the confidentiality of user-related data (note that data can be contained in a traffic channel, e.g., voice, or signaling channel, e.g., SMS messages).

The Subscriber Identity Module (SIM) is a security device, a smart card which contains all the necessary information and algorithms to authenticate the subscriber to the network. It is a removable module and may be used in any mobile equipment [32]. Note that the encryption algorithms are integrated into the mobile equipment as dedicated hardware. GSM does not use public-key cryptography. Symmetric keys are derived from user related data using an algorithm under the control of a master key.

The electronic payment system described in the example later in this paper, requires the SIM to contain a small payment application, based on the SIM Application Toolkit. The *SIM Application Toolkit* [13] provides mechanisms which allow applications, existing in the SIM, to interact and operate with any compliant mobile equipment. These mechanisms include displaying text from the SIM to the mobile phone, sending and receiving SMS messages, and initiating a dialogue with the user. In addition to the GSM security mechanisms, special SIM Application Toolkit security features have been defined [11, 12]. The security requirements that have been considered are: (entity) authentication, message integrity, replay detection and sequence integrity, proof of receipt and proof of execution, message confidentiality, and indication of the security mechanisms used. According to the standard, digital signatures can be used to implement some of these requirements.

Note that the same distinction between communications security and application security as made in the WWW security context, can be made here: standard GSM security at the communications level, and SIM Application Toolkit security at the application level.

3.2. WAP

The Wireless Application Protocol (WAP) is a protocol stack for wireless communication networks. WAP is bearer independent; the most common bearer is currently GSM.

Similar to SSL/TLS for the Internet, WTLS [39] is WAP's communications security solution. It also relies on a Public Key Infrastructure [35, 34]. The main differences are that WTLS supports by default algorithms based on elliptic-curve cryptography, is adapted for datagram communication (instead of connection), and supports its own certificate format, besides X.509v3, optimized for size. TLS was as such modified to make it more suitable in an environment where there are bandwidth, memory, and processing limitations.

At the application layer, WAP provides digital signature functionality through the WMLScript Crypto Library [40], which is similar to Netscape's Javascript signing. Comparable to the GSM's SIM, WAP devices will use a Wireless Identity Module (WIM) [38] which can contain the necessary private and public keys to perform digital signatures and certificate verification respectively.

3.3. Other systems and inititiatives

GSM is a second-generation system (2G). UMTS, Universal Mobile Telecommunications System [31], is part of a global family of third-generation (3G) mobile communications systems. These systems provide high-capacity and more secure [33] communication. A competitor of WAP is NTT DoCoMo's i-mode [23]. Bluetooth [5] is a wireless protocol for communication between devices that are in close proximity. The Internet itself is also expanding to the wireless world. The IETF is currently defining standards for Mobile IP [15], and is working on extensions (including wireless) for TLS [4].

The Mobile Electronic Signature Consortium has defined mSign [21], which should provide a standardized interface between Primary Service Providers (e.g., merchants) and Mobile Operators. It allows Primary Service Providers to request signatures from end-users through the Mobile Operators. The Mobile electronic Transactions initiative – MeT [22] – intends to establish a consistent and coherent framework for secure mobile transactions, based on existing standards and specifications; where needed, new functionality will be submitted to relevant standardization and specification organizations. There are numerous other fora concerned with mobile secure payments, see [7] for a description and comparison of these.

4. Combining WWW and wireless

Both the World Wide Web and the wireless world on their own have security and/or functionality problems. These shortcomings are explained in the following paragraphs. An approach in which the two worlds and their advantages are combined, is then motivated.

4.1. WWW: problems

It is very common that only web servers have certificates with which they are authenticated. In case user authentication is needed, it is almost never done via SSL/TLS client authentication. Users are often authenticated via their IP address, which is vulnerable to IP spoofing [3], which certainly does not provide mobility, and which is just not usable in an open system. Fixed passwords are frequently used, which provide mobility, but which are vulnerable to guessing, dictionary attacks and social engineering. Passwords that are only used once are not frequently used. They would be more secure, but certainly less convenient.

Root certificates are needed when verifying a web server certificate. It is very important that a user has an authentic copy of these certificates. This is more or less ensured by shipping them together with the browsers. It is however easy to add more or even replace root certificates. Moreover, the browser trust model causes a server certificate to be trusted if it is successfully verified by any of the root certificates (since there is usually no central policy management, this might easily include an attacker's root certificate). Finally, browsers generally also do not yet check if a certificate has been revoked.

If the user has a public/private key pair – for SSL/TLS client authentication, for SET, or for digitally signing documents – the private key will mostly reside on the hard disk of the machine. Even if it is protected by a pass phrase, it is still very vulnerable, for example due to Trojan horses. Users with such a software token are also hardly mobile. Smart cards are a solution, but for particular applications, they might be inconvenient. Moreover, smart card readers are currently not installed on each machine. Other special-purpose hardware, such as a Digipass [9], as sometimes used in e-banking, might be too costly for small applications, i.e., the investment for the customers and/or merchants would just be too high compared to the expected benefits.

Current end-user computing systems tend to offer more functionality at the cost of security. This is actually the reason why for example root certificates and private keys are so vulnerable on current end-user machines. Specifically, there is currently a lack of secure operating systems [19] and trusted components [30]. Today's PC and browser offer advanced functionality, but are therefore an insecure environment.

4.2. Wireless: problems

While the security problems on the WWW are currently more related to the secure management of the end-points, the security problems in some wireless systems are still with the protocols and algorithms themselves. For example, algorithms used by many GSM providers have been broken and 'over-the-air cloning' and real-time eavesdropping have been shown (at least in theory) to be feasible [28]. Security problems have been discovered in other mobile systems too [6, 17]. Most of these problems are due to non-public design of the algorithms and protocols, leakage and/or publication of the details to the general public afterwards, and discovery of flaws by the cryptographic community.

More conceptually, both GSM and WAP do not offer end-to-end security. GSM security only applies on the wireless link, i.e., from mobile phone to base station, but not from mobile phone to mobile phone. The fixed network is considered to be secure (more precisely, GSM intends to offer the same security level as the fixed network). In the WAP architecture, WAP devices communicate with web servers through a WAP gateway. WTLS is only used between the device and the gateway, while SSL/TLS can be used between the gateway and the server. From a security point of view, this means that the gateway should be considered as a person-in-the-middle. Note that WAP is now evolving into end-to-end security [37, 36].

Security seems to evolve in the good direction though. From a usability point of view on the other hand, mobile devices have still a rather limited functionality. They are not performant, and have often a quite poor human-device interface. Although mobile devices are getting more advanced, they will always be outsmarted by desktop PCs. Note that the complexity of the PC (e.g., multi-user operating system, data with executable content, ...) is the main reason why securing the end-points of the communication is such a difficult task, and remains an important problem on the WWW. As long as mobile devices stay quite simple and do not provide too much functionality, their security as an end-point will be more easy to cope with.

4.3. Motivation for a combined approach

By combining the World Wide Web with a wireless system, we want to come to practical and low-cost electronic commerce applications, which can fully exploit the broad functionality of the WWW. Two goals should hereby be achieved at the same time: *security* and *mobility*.

The WWW on its own does not seem to be sufficient for these applications. It surely provides broad functionality. When for example only fixed passwords are used, the WWW also offers mobility, i.e., a user can initiate transactions from any computer (e.g., a public terminal). Strong security is in that case however not achieved. Stronger security can be achieved by using for example cryptographic keys stored on the computer's hard disk. However, this does not allow for practical mobility. Special-purpose hardware tokens would increase the security of the application and provide mobility again. However, in an electronic commerce environment, consumers do not likely want to pay for a token that can only be used in the context of that application.

Wireless systems on their own are not suitable either. By definition, they offer mobility. Although there are some weaknesses in current systems, security in wireless systems tends to improve substantially. It is however clear that the GSM system is a rather limited environment. WAP offers a more general and WWW-like functionality, but in practice today's devices and networks do not satisfy the needs of merchants and customers. Mobile devices are generally expected to stay inferior to desktop computers.

This brings us to the motivation for a combined approach. Mobile devices are general-purpose devices which can be used as an extension to the WWW - instead of special-purpose devices - to offer more security and mobility without any extra cost. These mobile devices can be personalized and can store secret information such as cryptographic keys. They can be used in combination with any computer, i.e., the personal computer at the user's home, but also a public terminal, hereby providing mobility. Moreover, the computer terminal must not necessarily be completely trusted, as (part of) the security will rely on trusted and/or secret information that is securely stored in the device (and never leaves it, in case of secrecy).

In the remainder of this paper, this combined approach will be illustrated with an electronic payment system for the WWW that makes use of a mobile phone. This GSM based system is an alternative to the widely spread credit card based solution, offering more security and equivalent mobility and complexity (assuming that a mobile phone is standard equipment of many users). In addition, it might be suited for lower-price transactions.

5. GSM based payment for the WWW

The main goal of the remaining part of the paper is to present a system in which the WWW and GSM environment are combined to improve overall security, mobility, and functionality. In particular, an architecture and protocol are developed in which: (1) a customer can initiate and complete an electronic *payment* over the GSM network where the network operator is an active participant; (2) the *pre-payment* related interaction is done via the WWW; (3) the customer receives a receipt with which he/she can pick up the goods (*post-payment*).

5.1. Involved entities

The following entities play an active role in this e-commerce system:

Customer. The Customer wants to buy something via the WWW. Payment will be done via his/her GSM. The Customer will receive a receipt, with which he/she can pick up the goods (the system must work with both physically deliverable goods and electronically available goods). Obviously, the Customer should have a PC with Internet connection. This can also be a public terminal. He/she needs a mobile phone with SIM Application Toolkit functionality. The SIM card should be issued by a Network Operator that is running this electronic payment service. Optionally, there should be a connection between the mobile phone and the PC, and accordingly some extra software on the PC.

Merchant. The Merchant wants to sell something via the WWW. He/she should have a web server, and an access point to the mobile network. Examples are an on-line bookstore, a pizza delivery chain, an electronic parts shop, etc.

Deliverer. The Deliverer is the local (with respect to the Customer) representative of the Merchant. It will deliver the goods after having verified the receipt the Customer has obtained from the Merchant. The Deliverer should have some equipment to verify this receipt. An example is the pizza delivery boy/girl, etc. The Deliverer can also be another company that made an agreement with the Merchant. For example, the Merchant can send the goods to a gas station near the Customer; in this case, the gas station is the Deliverer where the Customer can pick up the goods.

Network Operator. The N.O. plays the role of the bank. It will deduct the necessary amount of money from the Customer's balance (can be credit or pre-payment based), and add this amount to the Merchant's balance. A commission on this amount will be taken, or a periodical fee will be requested from the Customer and/or Merchant. In practice there will be multiple N.O.s: N.O.(C), N.O.(M) and N.O.(D), for the Customer, the Merchant and the Deliverer respectively (as shown in Fig. 1).

Note that in reality, and from a non-technical point of view, it might not be easy for any Network Operator to deploy an electronic payment service (e.g., banking license). Alternatively, the "Network Operator" could in this system be replaced by a real financial institution, which makes an agreement with one or more operators.

5.2. Architecture and protocol

From a high-level point of view, the different entities perform the following interactions (see Fig. 1): after browsing and negotiating, the Customer requests a purchase; via an SMS message, the Merchant asks the Customer to pay the purchase; the Customer pays by sending an SMS message to the Network Operator; the Network Operator informs the Merchant about the successful payment; the Merchant sends a receipt to the Customer (also an SMS message); the Customer can use this receipt to pick up the goods at the Deliverer.

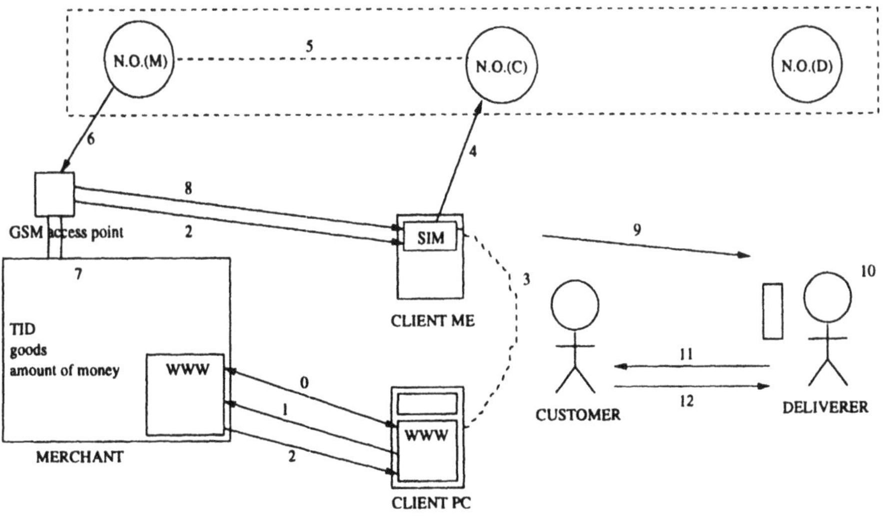

1. Purchase Request
2. Purchase Confirm
3. Verification by Customer
4. Debit Account
5. Inter-N.O.
6. Delivery OK
7. Verification by Merchant
8. Receipt
9. Presentation of Receipt
10. Verification by Deliverer
11. Delivery of Goods
12. Confirmation of Reception

Figure 1. GSM based payment for the WWW: architecture and protocol

The protocol contains the following steps (see Fig. 1):

1. Purchase Request. After browsing and negotiating (0), the Customer makes a *Purchase Request* via the WWW (1). The Merchant can choose the format and encoding of the message. It should at least contain a description of the goods, the amount of money to be paid, and the Customer's GSM number (in order to be able to send an SMS message to the Customer). The message will normally be sent through submission of an HTML form. The level of protection can be chosen by the Merchant, but it will normally be protected in transit by SSL/TLS. The form could also be digitally signed by the Customer (e.g., Netscape's Javascript signing capability, or Signed XML; note that a mobile device might in fact not provide any added value in this case).

2. Purchase Confirm. The Merchant sends a *Purchase Confirm* via SMS (2) to the Customer's mobile phone. This message should be in a standard format, and is optionally digitally signed by the Merchant. The message contains: (optionally) a description of the goods (either a hashed form of the description, or an abbreviated yet unique description of the goods, e.g., as in supermarket receipts), a Transaction ID (TID), a unique Merchant ID, the ID of N.O.(M), and the amount of money to be paid. The Merchant also sends a *Purchase Confirm* via the WWW (2). Note that this could already be included in the reply to the submission of the Purchase Request form.

3. Verification by the Customer. The Customer verifies whether all the ordered goods are listed, and whether the amount of money requested equals the amount agreed on. The information in the SMS message should be the same as the information displayed in the browser. Authentication of the Merchant thus relies on both GSM (we assume that the Customer knows the number of the Merchant) and SSL/TLS, so the Customer's trust in the correct execution of the transaction increases. If the reply in the browser and/or the SMS message are digitally signed, the signatures are verified. Note that in current GSM phones such a signature must possibly be verified using additional software on the computer. This requires a connection between the mobile phone and the PC which can for example be provided by Bluetooth. An automatic verification and comparison of the reply in the browser and the SMS message can then also be made. The interface to the Customer is provided by the SIM Application Toolkit. A payment application is installed on the SIM card, which is invoked on receipt of a Purchase Confirm message.

4. Debit Account. The SIM Application Toolkit application asks the Customer a confirmation for sending a *Debit Account* message (4) to the N.O.(C). This message includes the amount of money to be paid, the TID, the Merchant's ID and N.O.(M)'s ID. The authentication of the Customer relies on GSM entity authentication (the Customer's mobile phone number should be in the Merchant's database). The TID will allow verification by the Merchant afterwards.

5. Inter-N.O. The N.O.(C) deducts the proper amount of money from the Customer's balance, and forwards the Debit Account message to N.O.(M). The N.O.(M) adds the amount to the Merchant's account.

6. Delivery OK. The N.O.(M) sends a *Delivery OK* (6) to the Merchant. This message contains the amount of money and the TID, and can be digitally signed by the N.O.(M).

7. Verification by the Merchant. The Merchant verifies if the *Delivery OK* message originates from the N.O.(M) (relying on GSM entity authentication). If added, the digital signature of the N.O.(M) is verified. The Merchant looks up the TID in his transaction database, and checks if the amount of money is the same as included in the corresponding Purchase Confirm messages.

8. Receipt. The Merchant sends a *Receipt* (8) to the Customer via SMS. It contains: a (hashed) description of the goods, the TID, a timestamp (in order for the Deliverer to verify the freshness of the receipt), information on the Deliverer (optionally depending on the Customer's cell location, and including the Deliverer's GSM number), and information on the Customer (optionally including its GSM number, to allow verification of ownership of the receipt). The receipt is digitally signed by the Merchant. The receipt can only be used for the intended Deliverer as indicated. The TID and timestamp ensure that the receipt cannot be replayed by the Customer (i.e., the Deliverer should keep a list of previously received TIDs and should not accept receipts that are too old). GSM authentication is relied upon for authenticating the Customer.

9. Presentation of the receipt. If goods are electronic and delivered via the WWW, a receipt is not needed. Goods are then downloaded using the TID. The Merchant keeps a list of which TIDs correspond to transactions for which a payment has been received. Physical goods should be retrieved at the Deliverer. The receipt is forwarded to the Deliverer (9), manually or through the SIM Application Toolkit, or the Customer just presents the receipt to the Deliverer on the screen of his/her own GSM.

10. Verification by the Deliverer. The Deliverer just reads the receipt from the screen of the Customer's or his/her own GSM, or he/she verifies the receipt more properly by checking if the signature of the Merchant is valid. The Deliverer needs some infrastructure with GSM access point for this (e.g., a GSM connected to a laptop).

11. Delivery of goods. If the receipt is valid, the Deliverer can be sure that the Customer is the one that has made (and paid) the purchase. The goods can thus be delivered (11). In case of electronic goods which are delivered directly by the Merchant's web site (not necessarily though, as the Deliverer might have its own web site), the Customer should be granted access based on the TID: after a Delivery OK message has been received, the Merchant enables the access to the information; the TID should not be known to other entities (however, note that the N.O. should be trusted not to misuse its knowledge of the TID).

12. Confirmation of reception. After the Customer has obtained the goods, it can optionally be required that he/she confirms the reception of the goods (12), e.g., by digitally signing a specific message. This will prevent Customers from denying later on having received the goods.

6. Analysis and remarks

The proposed GSM based electronic payment system for the WWW is analyzed further in this section. Some GSM specific comments are given, the security and privacy of the system is evaluated, and a comparison with a number of similar systems is made. Note that this section only intends to discuss this particular example, and not the general combined approach.

6.1. GSM functionality

The protocol relies on SMS messages. These can only contain 160 characters, which should be taken into account when defining the exact content of the protocol messages. Note that GSM provides a mechanism to send long messages as a concatenation of multiple SMS messages. Since the protocol involves on-line bi-directional communication between the entities, there should be not much latency between sending and receiving SMS messages. This might be a problem in the case of international roaming.

6.2. Security

The security features of SSL/TLS and GSM form together a basis for the security of the proposed electronic payment system. By having a close link between the two, the security is even improved.

The Customer can securely request a purchase via SSL/TLS. The Customer will receive a confirmation via this same secure channel, and also on its mobile phone. Therefore, the Customer can double-check the Merchant's identity, and the contents of the purchase, including the amount of money to be paid.

The Merchant can rely on the GSM network to be sure to receive an authenticated payment from the Customer via the Network Operator later on. Moreover, the Customer cannot cheat by requesting its Network Operator to deduct a smaller amount of money than originally requested by the Merchant. The Merchant would notice the smaller amount of money and not send a receipt.

The Deliverer can validate a receipt by verifying the digital signature of the Merchant, and by checking if the receipt is fresh. Thus, receipts cannot be forged, and cannot be replayed. Moreover, if the Customer's mobile phone number is included in the receipt, the Deliverer could rely on GSM authentication and check if the receipt is actually presented by the original initiator of the transaction (provided that the Customer allows its own number to be sent to the receiving end; note also that for some applications, Customers might desire to be able to forward the receipt to another party that in its turn can pick up the goods).

As on top of SSL/TLS and GSM, some crucial messages are digitally signed; this decreases the need for Customers and Merchants to trust each other (i.e., they only need to trust they use the right public key, which should be ensured by the certificates that are issued by mutually trusted CAs). For example, since the receipt is digitally signed, it cannot only be verified by the Deliverer, but also by a Judge, in case of a dispute. Note that the latter also requires that the receipt includes a unique and indisputable description of the goods that should be delivered.

The Network Operator is trusted to transfer the proper amount of money from the Customer's to the Merchant's balance. It is expected to do so, as its business would otherwise quickly collapse due to negative publicity.

In some sense, the Customer's mobile phone can be considered as a secure and personal device (and care should therefore be taken that it is not easily stolen or lost). The strength of the electronic payment system proposed in this example relies particularly on the security of such a device, which is combined with the advanced yet insecure environment provided by the PC and the browser.

6.3. Privacy

The presented electronic payment system seems to offer more security than today's widely used mechanisms; however, it does not really offer more privacy. Merchants know at least the mobile phone number of their Customers. This number does not necessarily reveal a Customer's real identity (as opposed to an ordinary credit card payment). There already exist phone books with GSM numbers though. One would for example certainly not be happy when this number would be used for advertisement purposes. In fact, for this reason, some people will be reluctant to release their phone number, while they freely disclose their credit card number to merchants. The ability of hiding numbers or anonymizing customers in another way, would thus be an improvement of the system. Just as with credit card payments, the Network Operator knows exactly which Customers are buying goods from which Merchants and for what amount of money. The Network Operator will not necessarily know the actual nature of the goods though.

6.4. Other approaches

Other GSM based payment systems exist. GiSMo [14] is a system intended for the Internet in which customers receive a random code through SMS via a central server. This random code is then entered via the computer in order to pay. Mint [20] is a system in which each terminal/shop has a unique phone number which the customer should just call at the time of payment. Similar alternatives are Jalda [18] and Paybox [24].

In the system presented in this paper, more payment related information is exchanged via GSM, which results in a closer link between the WWW and the GSM interaction. Conceptually, it is also more general and independent of the wireless system. With more advanced mobile devices and networks, such as UMTS, more secure schemes would be possible, following the same architecture and protocol, but with different content of (and another exchange mechanism of) the messages. For example, instead of an account based protocol, electronic cash like schemes could be used. Mobile devices with built-in smart card readers would be very useful for integrating smart card based payment means as used in the physical world.

7. Conclusion

Electronic commerce is already a normal part of people's ordinary life. Mobile devices, and certainly mobile phones, are currently widely spread. This paper gave a brief overview of the security properties of the World Wide Web and some existing mobile systems. The main purpose of this paper was to suggest to use a wireless system as an extension to the WWW, to provide more security and functionality. To demonstrate this combined approach, a GSM based electronic payment for the WWW was presented.

Unlike most mobile phones, some mobile devices are powerful and advanced enough to allow more or less convenient browsing and shopping. Future mobile systems will also be more secure and will offer more functionality than the GSM system or than WAP. Yet, the concept of using an out-of-band channel for electronic payment, and the combined use of a mobile device together with a normal PC, will remain very useful.

For the PC and its big screen will always be far more advanced than the mobile device, but will never be mobile.

Acknowledgments

Joris Claessens is funded by a research grant of the Flemish Institute for the Promotion of Industrial Scientific and Technological Research (IWT). This work was also partially supported by the Concerted Research Action (GOA) Mefisto-666 of the Flemish Government.

This work was mainly initiated during a research visit at the Information Security Research Centre (ISRC) of Queensland University of Technology (QUT), Brisbane, Australia. The first author wants to thank Dr. Mark Looi of the ISRC for the interesting discussions on this topic. The authors also want to thank Dr. Silke Holtmanns of Ericsson for the various pointers to existing mobile payment systems, and the anonymous reviewers for their constructive remarks and suggestions.

References

[1] American Express. Private Payments. http://www.americanexpress.com/privatepayments/.

[2] Banxafe. http://www.banxafe.com/.

[3] S. M. Bellovin. Security Problems in the TCP/IP Protocol Suite. *Computer Communication Review*, 19(2):32–48, April 1989.

[4] Simon Blake-Wilson, Magnus Nystrom, David Hopwood, Jan Mikkelsen, and Tim Wright. TLS Extensions. IETF Internet Draft, June 2001.

[5] Bluetooth SIG. http://www.bluetooth.com/.

[6] Nikita Borisov, Ian Goldberg, and David Wagner. Intercepting Mobile Communications: The Insecurity of 802.11. http://www.isaac.cs.berkeley.edu/isaac/wep-draft.pdf.

[7] Clara Centeno. Mobile Payment Industry Fora - Consolidation of Initiatives Expected. *Electronic Payment Systems Observatory - Newsletter, ePSO-N*, (8):8–12, July 2001. Available at http://epso.jrc.es/.

[8] T. Dierks and C. Allen. The TLS Protocol Version 1.0. IETF Request for Comments, RFC 2246, January 1999.

[9] Digipass. http://www.vasco.com/.

[10] D. Eastlake, J. Reagle, and D. Solo. XML-Signature Syntax and Processing. IETF Request for Comments, RFC 3075, March 2001.

[11] ETSI. Digital cellular telecommunications system (Phase 2+); Security mechanisms for the SIM Application Toolkit; Stage 1. ETSI TS 101 180 (GSM 02.48).

[12] ETSI. Digital cellular telecommunications system (Phase 2+); Security mechanisms for the SIM Application Toolkit; Stage 2. ETSI TS 101 181 (GSM 03.48).

[13] ETSI. Digital cellular telecommunications system (Phase 2+); Specification of the SIM Application Toolkit for the Subscriber Identity Module - Mobile Equipment (SIM - ME) interface. ETSI TS 101 267 (GSM 11.14).

[14] GiSMo. http://www.gismo.net/.

[15] IETF Working Group. IP Routing for Wireless/Mobile Hosts (mobileip).

[16] InternetCash. http://www.internetcash.com/.

[17] Markus Jakobsson and Susanne Wetzel. Security Weaknesses in Bluetooth. In D. Naccache, editor, *Progress in Cryptology - Proceedings of the Cryptographers' Track at RSA 2001*, Lecture Notes in Computer Science, LNCS 2020, pages 176–191. Springer-Verlag, 2001.

[18] Jalda. http://www.jalda.com/.

[19] Peter A. Loscocco, Stephen D. Smalley, Patrick A. Muckelbauer, and Ruth C. Taylor. The Inevitability of Failure: The Flawed Assumption of Security in Modern Computing Environments. In *Proceedings of the 21st National Information Systems Security Conference*, pages 303–314, October 1998.

[20] Mint. http://www.mint.nu/.

[21] Mobile Electronic Signature Consortium. http://www.msign.org/.

[22] Mobile electronic Transactions. http://www.mobiletransaction.org/.

[23] NTT DoCoMo. i-mode. http://www.nttdocomo.co.jp/i/.

[24] Paybox. http://www.paybox.de/.

[25] Proton. http://www.protonworld.com/.

[26] E. Rescorla and A. Schiffman. The Secure HyperText Transfer Protocol. IETF Request for Comments, RFC 2660, August 1999.

[27] Eric Rescorla. *SSL and TLS: Designing and Building Secure Systems*. Addison-Wesley, 2000.

[28] Bruce Schneier. European Cellular Encryption Algorithms. *Crypto-Gram*, December 1999.

[29] SET Secure Electronic Transaction LLC. SET Secure Electronic Transaction Specification. http://www.setco.org/.

[30] Trusted Computing Platform Alliance. http://www.trustedpc.org/.

[31] The UMTS Forum. http://www.umts-forum.org/.

[32] Klaus Vedder. GSM: Security, Services and the SIM. In B. Preneel and V. Rijmen, editors, *State of the Art in Applied Cryptography*, Lecture Notes in Computer Science, LNCS 1528, pages 227–243. Springer-Verlag, 1998.

[33] M. Walker. On the security of 3GPP networks. Invited talk at Eurocrypt 2000.

[34] Wireless Application Protocol Forum. WAP Certificate and CRL Profiles. Approved 22-May-2001.

[35] Wireless Application Protocol Forum. WAP Public Key Infrastructure. Version 24-Apr-2001.

[36] Wireless Application Protocol Forum. WAP TLS Profile and Tunneling. Version 11-April-2001.

[37] Wireless Application Protocol Forum. WAP Transport Layer End-to-end Security. Approved Version 28-June-2001.

[38] Wireless Application Protocol Forum. WAP Wireless Identity Module, Part: Security. Version 12-July-2001.

[39] Wireless Application Protocol Forum. WAP Wireless Transport Layer Security. Version 06-Apr-2001.

[40] Wireless Application Protocol Forum. WAP WMLScript Crypto Library. Version 20-Jun-2001.

ON MOBILE AGENT BASED TRANSACTIONS IN MODERATELY HOSTILE ENVIRONMENTS *

Niklas Borselius, Chris J. Mitchell, Aaron Wilson
Mobile VCE Research Group, Information Security Group,
Royal Holloway , University of London
Egham, Surrey TW20 OEX, UK
Niklas.Borselius@rhul.ac.uk, C.Mitchell@rhul.ac.uk, aaron@gmx.co.uk

Abstract When using mobile agents, numerous security issues must be considered. In this note we propose two methods to improve the security and reliability of mobile agent based transactions in an environment which may contain some malicious hosts.

Keywords: mobile agent, digital signature, transaction, security

1. Introduction

In this paper we consider strategies for the deployment of mobile trading agents to reduce certain security threats to their operation. In a future world of co-operating mobile and fixed devices, the mobile agent computing model is expected to become an increasingly important one. In the domain of e-commerce/m-commerce transactions, mobile trading agents could play a very useful role. Users could launch such agents to make transactions on their behalf, and the agents would look for the 'best buy' by visiting multiple merchant sites without any direct user intervention. Indeed such activity could take place while the user has no current network connectivity.

The mobile agent computing model gives rise to a range of security threats. These threats can be divided into two main classes:

*The work reported in this paper has formed part of the Software Based Systems work area of the Core 2 Research Programme of the Virtual Centre of Excellence in Mobile & Personal Communications, Mobile VCE, www.mobilevce.co.uk, whose funding support, including that of EPSRC, is gratefully acknowledged. More detailed technical reports on this research are available to Industrial Members of Mobile VCE.

- threats to the platform from malicious and/or unauthorised agents, including threats to the integrity of the platform and other agents, threats to the confidentiality of stored data, and denial of service threats, and
- threats to the agent from malicious platforms, including threats to the confidentiality of agent stored data, and threats to the integrity of the agent and its computations.

In this paper we are concerned with the second class of threats, and in particular with threats to agents deployed for trading applications. Specifically, users will need to give trading agents certain authority to authorise transactions, whilst at the same time users will wish to protect themselves against malicious merchants forcing an agent to make a non-optimal purchase.

We consider simple ways in which deployment of multiple agents can reduce the threat to trading agents from platforms outside of their direct control. We consider two general approaches. In the first approach multiple agents are equipped with 'shares' of the means to commit to a transaction. In the second approach a single trusted host provides a location for multiple agents to 'report back' information enabling a purchasing decision to be made.

The paper has the following structure. The next section explores threats to trading agents in more detail. This is followed in Sections 3 and 4 by a discussion of the models used here for agent platforms and for trading agents. Sections 5 and 6 then explore the two approaches to enhancing trading agent security.

2. Agent Security Issues

The use of mobile agents raises a number of security concerns. Agents need protection from other agents and from the hosts on which they execute. Similarly, hosts need to be protected from agents and from any party which can communicate with the platform. The problems associated with the protection of hosts from malicious code are quite well understood.

The problem of malicious hosts seems the hardest to solve. In fact some people hold the opinion that it is insoluble. The particular attacks that a malicious host can make have been described in [Hoh98a] and [Has00], and can be summarised as follows.

- Observation of code, data and flow control,
- Manipulation of code, data and flow control – including manipulating the route of an agent,

- Incorrect execution of code – including re-execution,
- Denial of Execution – either in part or whole,
- Masquerading as a different host,
- Eavesdropping of agent communication,
- Manipulation of agent communication,
- False system call return values.

There have been many attempts to address these threats either completely or in part. Most of these attempts fall into one of the following broad categories.

- The first category comprises approaches that do not allow an agent to leave a trusted environment. Solutions to this include using a host infrastructure that is operated by a single party, allowing agents to migrate only to trusted hosts [FGS96], or possibly hosts with a good reputation [RJ96].
- The second category is pragmatic; it consists of solutions to a single part of the malicious host problem. These consist of agents detecting when they have been modified [Vig97], and proof verification techniques [Yee97].
- The third class consists of assuming that there is special, tamper-proof hardware available, see for example [Yee97] or [WSB98].
- The final category uses software methods to obscure the code from the host. Approaches include obfuscation [Hoh98b] [Ng00], mobile cryptography [ST98, ST97] and using environmental conditions to hide parts of the code [RS98].

The approaches described in this paper, based on replicating agents, do not fit into any of the above four classes. There appears to be relatively little literature devoted to this approach to dealing with the threats to agent security.

We now consider the threats to a trading agent in more detail.

2.1. Threats to trading agents

We now turn to look at the particular threats to an agent which wishes to purchase an item (or a service) from a merchant. These all fall into the categories above. We concentrate on the threats to an agent involved in a trade, rather than more general threats.

1 A malicious host lies about offer.

Here a host lies about the offer it makes to an agent, in order to get the trade. The host would then charge a higher price at a later date. One way around this is to force the host to sign its bid, thereby committing to it.

2 A malicious host learns other offers and undercuts them.

If a host knows that all offers but its own have been collected and finds out the best standing offer, it can undercut the best standing offer slightly (in fact the host need not know all other offers, it could just undercut the current offers). (In some circumstances letting hosts undercut each other might be considered a desirable feature.)

3 A malicious host learns the price a user is prepared to pay and bids just under this.

In a similar fashion the host may charge more than its normal price, if it knows the maximum price the user is prepared to pay. Thus a host must be kept from learning the maximum price a user is prepared to pay, either by encrypting this information or by not sending this information with the agent.

4 A malicious host manipulates the requirements.

This is when the host changes the requirements to favour its bid. For example, it could add a requirement to buy from a certain host, or remove constraints from the agent.

5 A malicious host alters the agents route.

Here, the host keeps the agent away from its competitors, and thus secures the agent's trade. One way to prevent this is to use more than one agent (possibly an agent per host), send each agent on a different route and combine the offers on the agent's return. Another way is to use one agent with a 'star' like route – it returns home after visiting each host before being sent out to a different host.

6 A malicious host commits to purchases that the user does not wish to make.

This happens when a host can abuse the committal function that an agent has. A method to discourage this is to force the host to sign a transaction, as well as the user (thus providing traceability).

7 A malicious host denies the agent a service.

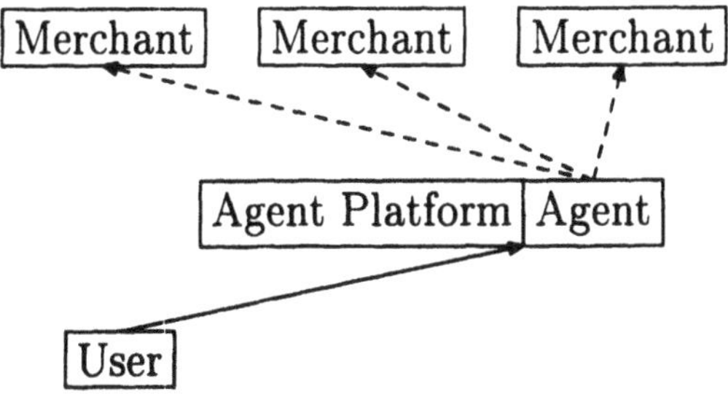

Figure 1. A model for agent platforms

Here a host would stop an agent from moving further on its route. This of course could be traced if an agent reports when it arrives on a host.

8 A malicious host captures electronic money.

Here a host would remove the electronic money that an agent may have to purchase an item and either steal the money outright, or use it for a different purchase.

We do not consider the payment process here, as we are concerned only with the part of a transaction involved in selecting a merchant and committing to the transaction.

3. Models of Agent Platforms

Mobile agents roam between platforms. However, they can also communicate with each other, and with other hosts. This leads to the question as to the best "platform" model to use for trading (or indeed any other) agents. There are clearly two basic approaches which we now describe.

The first approach (see Figure 1) is to have a designated platform (or a collection of such platforms) to which we can send an agent to execute. This agent then communicates with merchant servers to seek information and commit to purchases.

The second model (see Figure 2) is to have an agent roam to each merchant server in turn and collect the information it requires. After collecting all the information the agent can then either return to the user to make the purchase, return to the chosen merchant to make the purchase or make the purchase from the final host.

In a mobile telecommunications environment it may also be beneficial to have a third model. This is where the requirements for a purchase are communicated to a 'home platform' (the user's home PC or a network operator controlled device) which then forms the agent and conforms to one of the above models.

In the above, any of the platforms may be malicious, with the possible exception of the home platform. The solutions proposed below can be made to fit into any of the above situations, although they both fit better into the first model.

The security risks associated with the above two models clearly differ. In the first case, the 'designated platform' might be trusted to keep secret certain agent information. An example of where this might be useful is when the agent contains details of the user 'expected' price (or maximum price), which it would be helpful not to reveal to the merchant. Of course, the threat then arises that one of the designated platforms will collude with one or more of the merchants. In the second case, it is clearly impossible to try and keep any information in the agent secret from the merchants. In both cases, however, as we will show in the remainder of this paper, there are potential benefits to be gained from the use of multiple agents, albeit not from the confidentiality perspective.

4. Model for a trading agent

We consider the information that an agent wishing to trade must know. Firstly, when initiating a purchase, a user will have a set of requirements (for instance the item to be purchased, the maximum price for that item, a time limit within which the purchase to be made). We will assume that a user encodes these requirements into a string R which is understood by all parties. When a server quotes for a given purchase, it will also produce a similar string with its offer.

The agent, if it is to perform the purchase on behalf of the user, must also carry a function which will commit to the trade. This could be performed by, for instance, signing the details of the trade. One scheme to allow an agent to perform a signature operation on behalf of a user without revealing the user's private key to a host is proposed in [KBC00].

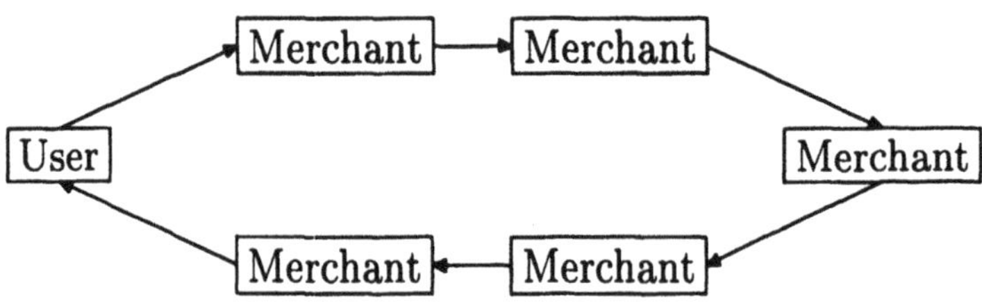

Figure 2. A second model for agent platforms

In this scheme, using RSA, an agent carries both the hash value, h, of the requirements and the signed hash value, $h^d \bmod n$, where (d, n) is the user's private RSA key. To commit to a transaction for the user the agent calculates

$$(h^d)^x = h^{xd} = (h^x)^d \bmod n$$

effectively signing h^x where x is the server's offer. An alternative to this where the agent carries its own private key which the user certifies is given in [BMW01a].

Thus we assume that a trading agent will carry the following information:

- User Identifier – U

- Requirements for purchase – R

- A committal function – $\mathcal{C}$. The committal function is used by the agent to commit to a transaction on the user's behalf. $\mathcal{C}$ could be a signature function using a special private key provided to the agent by the user. Alternatively, $\mathcal{C}$ could be a function of the type described above, derived from the user's own private signature key. In any event, we assume that the function is designed so that only transactions within constraints defined by the user can be authorised.

Note that, if a single 'trading agent' is deployed there are a number of problems which might arise. Firstly, although the committal function will typically be limited to transactions conforming to user-defined parameters, there is still the possibility that the agent platform will force the agent to commit to a transaction which is less than optimal. It may also commit to more than one transaction, even if the user only intended to make at most one purchase.

One way to reduce this threat is to deploy multiple agents, a subset of which must agree to the transaction before it can be authorised. Such an approach is the focus of the remainder of this paper.

5. Threshold Scheme

We attempt to solve the malicious host problem by using multiple agents each of which has a 'vote.' If one of the possible transactions receives enough votes, then a transaction will be authorised with the relevant merchant. We begin by outlining the scheme, and then consider the details of what a secure vote can consist. We assume use of a (k, n) scheme – *i.e.* a server will need k votes out of a possible n to 'win'.

5.1. The Scheme

Let $T = \{T_1, T_2, \ldots, T_n\}$ be a set of agent platforms. The user then sets up a (k, n) voting scheme with shares $s_1, s_2, \ldots, s_n$. Clearly k should exceed the number of 'suspected' malicious hosts. Given no information about the system a sensible value would probably be $n/2+1$. The value of k reflects the level of trust in the system.

The user then forms n agents A_i $(1 \leq i \leq n)$ containing the following information

- User Identifier – U
- Requirements for purchase – R
- A vote – s_i

Each agent is then dispatched to its agent platform. At the platform there are two modes of execution:

1 The agent contacts each merchant itself, and gathers bids that meet the requirements.

2 The agent contacts a subset of the merchants and communicates the best bid to its peers.

We note that case (2), unless only contacting a single host, is a situation that must be carefully thought out. This is because if there is no overlap in the merchants, collusion may mean that attacking less than k servers is necessary.

When each agent has received all the information about each bid, the agent sends its vote to the merchant with the best offer. On receipt of the correct number of votes, the merchant or a nominated third party can construct (and verify) from the votes the authorisation for the bid. The merchant or nominated third party can then use this as evidence that the user has committed to transaction.

We now consider the security of the above scheme. The major advantage of the scheme is the need to corrupt either $n-k+1$ agents to prevent the transaction or k hosts to divert or alter the transaction. Thus the choice of k is crucial.

This also means that a denial of service attack is harder as a server or set of colluding servers will need to terminate (or prevent from communicating their vote) $(n-k)+1$ agents. Again to force a purchase a host or hosts must force k agents to offer their vote.

If an agent visits a subset of the servers involved, the information could then be used to help identify any malicious hosts.

5.2. The Votes

As mentioned above, the votes can be assembled by either the selected merchant or a nominated third party. Note that there are clear risks associated with giving votes to the merchant, since the merchant could now possibly commit the user to a transaction of the merchant's choice (within any constraints imposed by the string R). That is, the merchant is not forced to commit to the transaction as offered to the agents. Hence the use of a nominated third party to reconstruct the votes is the preferred approach. The possibility that this may not be feasible in practise leads to an alternative approach.

One approach is *threshold cryptography.* Threshold cryptography was first proposed by Desmedt [Des88]. A typical example of a threshold cryptosystem is one that would allow a set of t parties to sign any document such that any coalition of less than t parties cannot sign any other document. Schemes tend to rely on a combiner which does not necessarily need to be trusted. Schemes based on both RSA and El Gamal have been proposed.

Recently Shoup [Sho00] proposed an RSA scheme which is as efficient as possible; the scheme uses only one level of secret sharing, each server sends a single part signature to a combiner and must do work that is equivalent, up to a constant factor, to computing a single RSA signature. Although not perfect as a threshold signature scheme (as it relies on a trusted party to form the shares) this scheme is ideal in our setting. (Note that an alternative scheme without a trusted dealer is given in [DK01]. This scheme also improves on Shoup's scheme by not relying on an RSA modulus made up of 'safe primes'). An example of an El Gamal scheme is given in [Lan95]. We note that a (n, n) threshold signature scheme is just a multisignature; such schemes have been studied for many years — see, for example, page 488 of [MvOV96].

We note, however, that such a threshold signature scheme does not provide a means for the shares to incorporate an encoding of the string R. Thus, if there were k colluding hosts they could sign (and reconstruct a signature) for any document. One solution to this problem is for the user to generate a special signature key pair for the particular purchase (i.e. for this particular set of agents), and then to generate a certificate for the public key incorporating a copy of R. When the signature is reconstructed from the signature shares, it can be verified using this certificate. However, it is possible to merge the undetachable signature scheme given in [KBC00] with the threshold signature scheme of Shoup [Sho00] and details of this are given in [BMW01b].

6. Using one trusted host

We consider a second solution to the problem, which employs a single trusted host. We note that the solution described below involves a user sending out agent(s) to individual merchant servers, whereas it could just communicate with them to ask for their bids. However, in a wireless communications setting where communication is expensive, slow, and/or unreliable, it is believed to be beneficial to be able to dispatch an agent into the fixed network. When the agent has finished its task it contacts the user or waits for the user to collect the result.

Let $S = \{S_1, S_2, \ldots, S_n\}$ be a collection of servers offering a service that a user wishes to purchase. Let T be a host that the user trusts to act honestly in this transaction. (Note that we do not need to trust this host fully – it just needs to be neutral in this transaction). Before the transaction commences we assume that each server S_i and T securely establishes a shared secret key K_i. Optionally, a key for message integrity checks could also be established.

The user despatches an agent A to the trusted host containing the information outlined in §4. We note that the committal function $\mathcal{C}$ may be of any form with which the user is prepared to trust the host T. However, to reduce the trust requirements we envisage that this will be the scheme outlined in either [BMW01a] or [KBC00].

There are now several approaches for T. The first is to form a single subagent containing the following information

- Agent identifier – I
- Requirements for purchase – R
- Host identifier – T

which would then visit each of the servers in S in turn. We note that the requirements sent out do not need to include pricing information (that is the maximum price the user is prepared to pay) or any other information that the user wishes to be used to help make the decision, but does not wish to communicate to the server. Another approach is to form a single agent for each server. A third approach has the above agent visiting a subset $S' \subset S$ of the above servers. Whichever strategy is employed, at each host the agent performs the following actions:

1. Find out the server's bid B_i for the item specified in the requirements R.

2. Encrypts the concatenation of B_i, R, S_i and I using either K_i. At this point the server could also, optionally, attach a symmetric

MAC (Message Authentication Code) to the bid to protect the integrity of the server's bid. Label the encrypted string E_i.

3 The agent then stores the pair (S_i, E_i).

The agent returns to T when it has finished visiting all of its servers. The agent on T then decides the best offer and commits to it using the committal function.

We note some of the features of the above scheme.

- Using an agent per server really alleviates the need to encrypt anything, assuming that agents are always transferred between hosts in encrypted form.
- Using a single agent leaves yourself open to some attacks.
- Using more than one agent that does not visit all the hosts could then be used to (help) identify a malicious host.

If we use a single agent and it visits all the hosts, or we have an agent that visits more than one host, the agent is subject to the following attacks:

- An approach to enable a malicious host to underbid its competitors, is as follows. The host forms a new agent containing the user's requirements, a fictitious user identifier, and its own host identifier. This agent would then traverse the route of the user's agent, and discover the bids offered for that set of requirements. The host could then under bid its competitors, but the user's agent would have had to have been kept on the malicious host in the interim period. Thus monitoring the progress of an agent could help determine if such an attack was being used.
- A simple denial of service attack: stop the agent in its tracks. If there is no progress monitoring (*e.g.* agent at host S_i) then this attack is hard to defeat.
- A malicious host could alter the pair (S_i, E_i) to read $(S_j, junk)$ (where *junk* is a random string of the correct length) to stop the decryption of a bid. However as the host cannot read the bid, for this to be successful (*i.e.* to delete those bids more attractive than those of the malicious host) the host would have to have knowledge of all the bids - which it would have to gather itself (possibly by cloning the agent).

If we use an agent that visits a subset of the hosts, and assume that the malicious host already knows the "best offer" at any given point, it will then try to undercut this (this undercut is a lie). If we then require, we can apply rules to the results of the other agents, and attempt to identify the malicious host. This also requires careful choice of agent destinations and routing.

Note that to force T to purchase from a malicious host, the host has to lie and then be unscrupulous, or just lie and possibly not profit as much as it would expect. That is if the malicious host M wants to force a user to trade with it, then it must have the best price. So it must either charge more than its advertised price (possibly breaking the committal function) or make less profit than it expects (because the price advertised is less than the host should sell for).

We now consider the extent to which the user must trust the host T. The user must trust that T does not favour a particular server for this transaction. However, with a sufficiently good committal function then this is the only trust requirement. For example using the Kotzanikolaou et al. undetachable signature scheme [KBC00], as a committal function, T can be given the means to commit to the transaction without being trusted with a copy of the user's private signature key. This may be a situation where using an undetachable signature scheme has advantages over the creation of a separate signature key for each agent.

7. Conclusions

We have considered two different ways in which the deployment of multiple agents can reduce the threat to trading agents from potentially malicious agent platforms. In the first approach multiple agents are equipped with 'shares' of the means to commit to a transaction. A method implementing this idea using a threshold signature scheme, e.g. the recently proposed scheme of Shoup, [Sho00], was outlined. In the second approach a single trusted host is employed to collect information from multiple agents on possible transactions. This host then chooses the optimal transaction and commits to it.

The two approaches each have their own advantages. The first approach avoids the need for a single trusted host. However, implementing the first approach requires use of some potentially complex cryptographic signature functions. The second approach is potentially less complex from a cryptographic perspective, but does require a host which, if not completely trusted, is at least required to act neutrally with respect to the set of merchants. Both approaches are of potential practical importance in future mobile computing environments.

References

[BMW01a] Niklas Borselius, Chris J. Mitchell, and Aaron Wilson. A pragmatic alternative to undetachable signatures. Preprint, 2001.

[BMW01b] Niklas Borselius, Chris J. Mitchell, and Aaron Wilson. Undetachable threshold signatures. To be presented at the IMA Conference on Cryptography and Coding, December 2001 (proceedings to be published in the Springer-Verlag LNCS series), 2001.

[Des88] Y. Desmedt. Society and group oriented cryptography. In C. Pomerance, editor, *Advances in Cryptology - Crypto '87 proceedings*, number 293 in LNCS, pages 120–127. Springer-Verlag, Berlin, 1988.

[DK01] Ivan Damgård and Maciej Koprowski. Practical threshold RSA signatures without a trusted dealer. In Birgit Pfitzmann, editor, *Advances in Cryptology - EUROCRYPT 2001*, number 2045 in LNCS, pages 152–165. Springer-Verlag, Berlin, 2001.

[FGS96] William Farmer, Joshua Guttmann, and Vipin Swarup. Security for mobile agents: Authentication and state appraisal. In *Proceedings of the European Symposium on Research in Computer Security (ESORICS)*, number 1146 in LNCS, pages 118–130. Springer-Verlag, Berlin, 1996.

[Has00] Vesna Hassler. *Security Fundamentals for E-commerce*. Artech House, 2000.

[Hoh98a] Fritz Hohl. A model of attacks of malicious hosts against mobile agents. In *Proceedings of the ECOOP Workshop on Distributed Object Security and 4th Workshop on Mobile Object Systems: Secure Internet Mobile Computations*, pages 105–120, 1998.

[Hoh98b] Fritz Hohl. Time limited blackbox security: Protecting mobile agents from malicious hosts. In Giovanni Vigna, editor, *Mobile Agents and Security*, number 1419 in LNCS, pages 92–113. Springer-Verlag, Berlin, 1998.

[KBC00] Panayiotis Kotzanikolaou, Mike Burmester, and Vassilios Chrissikopoulos. Secure transactions with mobile agents in hostile environments. In E. Dawson, A. Clark, and C. Boyd, editors, *Information Security and Privacy, Proceedings of the 5th Australasian Conference ACISP 2000*, number 1841 in LNCS, pages 289–297. Springer-Verlag, Berlin, 2000.

[Lan95] Susan K. Langford. Threshold DSS signatures without a trusted party. In D. Coppersmith, editor, *Advances in Cryptology - Crypto '95 proceedings*, number 963 in LNCS, pages 397–409. Springer-Verlag, Berlin, 1995.

[MvOV96] A. Menezes, P. van Oorschot, and S. Vanstone. *Handbook of Applied Cryptography*. Discrete Mathematics and Its Applications. CRC Press, October 1996. Available on-line at http://www.cacr.math.uwaterloo.ca/hac.

[Ng00] Sau-Koon Ng. Protecting mobile agents against malicious hosts. Master's thesis, The Chinese University of Hong Kong, June 2000.

[RJ96] Lars Rasmusson and Sverker Jansson. Simulated social control for secure internet commerce. In *New Security Paradigms '96*, pages 18–26. ACM Press, 1996.

[RS98] James Riordan and Bruce Schneier. Environmental key generation towards clueless agents. In G. Vigna, editor, *Mobile Agents and Security*, volume 1419 of *LNCS*, pages 15–24. Springer-Verlag, Berlin, 1998.

[Sho00] Victor Shoup. Practical threshold signatures. In Bart Preneel, editor, *Proceedings of EuroCrypt 2000*, number 1807 in LNCS, pages 207 -220. Springer-Verlag, Berlin, 2000.

[ST97] Tomas Sander and Christian Tschüdin. Towards mobile cryptography. Technical Report 97-049, International Computer Science Institute, Berkeley, 1997. Available at http://ww.icsi.berkeley.edu/ sander/publications/tr-97-049.ps.

[ST98] Tomas Sander and Christian Tschudin. Protecting mobile agents against malicious hosts. In Giovanni Vigna, editor, *Mobile Agents and Security*, number 1419 in LNCS, pages 44–60. Springer-Verlag, Berlin, 1998. Available from http://www.icsi.berkley.edu/ sander/publications/MA-protect.ps.

[Vig97] Giovanni Vigna. Protecting mobile agents through tracing. In *Proceedings of the Third ECOOP Workshop on Operating System support for Mobile Object Systems*, 1997.

[WSB98] U. G. Wilhelm, S. Staamann, and L. Buttyán. On the problem of trust in mobile agent systems. Available from http://www.isoc.org/isoc/conferences/ndss/98/ndss98.htm, 1998. Network and Distributed System Security (NDSS'98) Symposium.

[Yee97] Bennet Yee. A sanctuary for mobile agents. In *DARPA Workshop on Foundations for Secure Mobile code*, 1997. Available from http://www.cs.nps.navy.mil/research/languages/statemensts/bsy.ps.

SPARTA

A Mobile Agent based Intrusion Detection System

Christopher Krügel*
Distributed Systems Group, Technical University Vienna
chris@infosys.tuwien.ac.at

Thomas Toth
Distributed Systems Group, Technical University Vienna
ttoth@infosys.tuwien.ac.at

Engin Kirda
Distributed Systems Group, Technical University Vienna
ek@infosys.tuwien.ac.at

Abstract The large number of machines with different operating systems and applications in an enterprise network makes it very difficult for the system administrator to close all security holes and install the latest OS and software patches. When the network is connected to the Internet and services are remotely available they become a potential target for hackers. As the number of security related incidents is constantly increasing at an alarming rate the need for automated tools to detect intrusions becomes evident. Such tools are called intrusion detection systems.

We present Sparta, a system that allows to detect security policy violations and network intrusions in a heterogeneous, networked environment. We have designed a pattern language in order to express intrusions (i.e. offending event patterns) in a declarative manner. This allows to specify what to detect instead of how to detect. A fully distributed approach to find the given patterns is presented as well. We use mobile agents to correlate event data instead of moving the whole information to a central location. This increases the fault tolerance and scalability of our system.

Keywords: Intrusion Detection, Mobile Agents, Pattern Specification Language, Event Correlation, Network Security

*Contact Author

Introduction

Virtually every organization depends on sensitive data which has to be protected against unauthorized access. Such data is often stored on machines which are remotely available over a network. The growth of the Internet has caused an increase of the size of individual networks as well as an increase of transported traffic. This makes it extremely difficult to manually manage and protect valuable assets. Combined with an alarming rise of attacks and hacking attempts, organizations need tools like intrusion detection systems (IDS) to enforce security and detect hacking attempts.

Sparta (an acronym for Security Policy Adaptation Reinforced Through Agents) is the name of a system architecture which is capable of monitoring a network to detect network intrusions and security policy violations. The system monitors local events at hosts which are connected by a network, relates them and provides an interface where the user can query the gathered information. This makes it possible to apply our design to a broad range of applications and use it for a number of network related tasks, ranging from network management to intrusion detection.

The contribution of this paper is the description of an architecture to collect and relate distributed data in an efficient way by using mobile agents and its application to network intrusion detection. In contrast to traditional designs where data is gathered and analyzed at a central location, the application of mobile agents allows distributed analysis. This approach improves the scalability and increases the fault tolerance in our opinion.

1. Functional Description

Sparta is an architectural framework which allows to identify and relate interesting events that may occur at different hosts on a network. A single event is described by specifying appropriate values for its attributes. A number of events can be connected by defining temporal or spatial relationships between them or imposing certain constraints on their attributes thereby creating a pattern. In order to deal with complex patterns and systems, it is not sufficient to select events based on content alone. It is necessary to consider multiple events at the same time and deduce knowledge that is beyond the scope of an individual event. The process of detecting a set of events with given properties is called *correlation*.

This general correlation capability allows the Sparta architecture to be used for different distributed applications, ranging from network security to network management implementations. We currently build a security policy and ID application based on our design.

The basic functionality can be described as follows. Interesting events are locally collected and stored. The collection of all local information can be considered as a distributed database with horizontal fragmentation. For each relation (i.e. event type), the tuples (i.e. actual events) are stored at different locations. A user may issue queries in our Event Query Language (EQL) to search for a set of events that fulfill his desired constraints. In addition to this the system can also be used to gather statistical information. It is possible to find the number of pattern instances at each host and to calculate the maximum or minimum for event attribute values as well as their sums over a set of hosts. The query is carried out by mobile agents which return their results to the user.

For our intrusion detection system, a failed authentication attempt or the start of a root shell might be examples of interesting events. Sparta allows to count the number of failed telnet logins for a certain user throughout the network (to detect distributed door knob rattling attempts) or to find tree-like connection patterns between hosts (to identify a spreading worm). It is important to notice that event correlation might yield information that is impossible to gain by just looking at a single node. Consider an intruder who tries to cover his tracks by performing several consecutive telnet logins (i.e. producing a **telnet chain**). This is an often observable behavior that exploits the fact that different machines are administered by different people and don't have synchronized local clocks. Tracing an attacker by having to look at all these logfiles is rather difficult. On each local machine only a simple incoming and outgoing connection is noticed but when looking at the entire network the offending pattern becomes evident. GrIDS (Staniford-Chen et al., 1996) is a well known ID system which bases its detection solely on looking for connection patterns but uses a different mechanism to collect and relate data.

2. System Architecture

The system consists of a set of hosts connected by a network where each node has the following components installed (see Figure 1).

- Local event generator (sensor)
- Event storage component
- Mobile agent platform
- Agent launch and query unit (optional)

The local event generation is done by sensors which monitor interesting occurrences on the network (network based) or at the host itself (host based detection). The exact types of events and their attributes as well as the implementation of the sensors are mainly determined by the application's needs. The type of an event is represented by the type of the class in the implementation (i.e. Java class), with the event's attributes being stored by the members of the corresponding class. It is possible to extend an event by subclassing from an existing one and add the desired additional information. This allows to

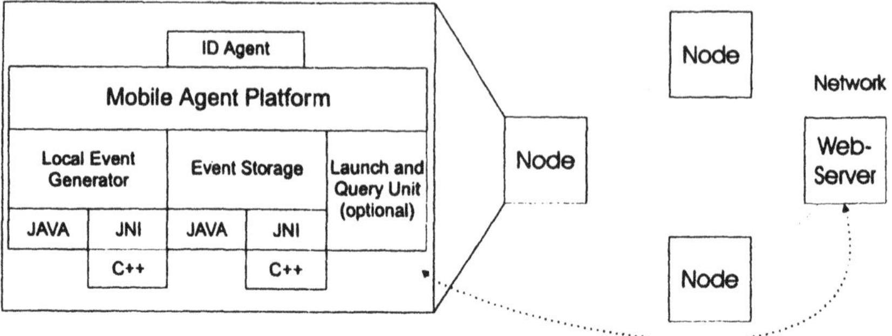

Figure 1. Sparta Architecture

write patterns which relate high level events and have the system automatically consider all actual instances (i.e. subclasses) of such generic events.

Sensors store their generated data in a local data storage component, preferably a database. The data storage component must be able to support the inheritance relationship of events. When queries specify parent class events, derived events have to be returned as well.

The mobile agent subsystem is responsible for providing a communication system to move the state and the code of agents between different hosts and for providing an execution environment for them. Additionally, the system has to provide protection against security risks involved when utilizing mobile code (see Section 5 for more details). An important task of the agent subsystem is the provision of a directory service. When agents have to look for event patterns, they need to access a list of all hosts with an installed agent platform. The agent platform also provides clock synchronization with a maximum guaranteed deviation. This is needed to be able to temporally relate events at different nodes.

The user interface allows users to specify queries and claim the results. The agent launch and query unit initiates the launch of appropriate agents and provides a way for them to communicate back their results. Queries are written in a language called Event Query Language (EQL), which we have developed to conveniently specify patterns that reflect a security violation. This is described in more detail below in Section 3.1. The user interface itself is realized as a web interface using HTML and JavaScript on the client side and Servlets on the server side. The communication between the client and the server is secured by using SSL connections. This setup allows a user to access the system via a standard browser from any computer that needs no Sparta components installed.

3. Pattern Specification

The design of our pattern specification language is guided by two conflicting goals. The first goal states that the language should be as expressive as possible.

It would be desirable to allow the description of complex relationships between events on different hosts using regular or tree grammars. Unfortunately, the evaluation of complex patterns makes it necessary for each local host to send a huge amount of data to a central site. This conflicts with the second goal, which demands that the amount of data that has to be transferred between hosts should be as small as possible. When a system uses mobile code (i.e. mobile agents), it should aim at performing flexible computation remotely at the location where the interesting data is stored instead of abusing agents as simple data containers.

When the interesting patterns do not change frequently, it would be desirable to wire them directly into local components at each host. For our application, users intend to specify many different patterns and perform a lot of ad-hoc queries. Therefore, the application of mobile code is reasonable.

The basic building block of a pattern is a set of local events. One can specify a list of events on a local host by enumerating them and imposing certain constraints on their attributes. A constraint can have two different formats. One format allows to relate an event attribute with a constant value, using one of the standard logical operators or one of our self-defined ones (`in`, `range`). The other format allows to relate an attribute of one event with another attribute of the same or a different event, again using the full range of operators. This allows to select a number of events with a common context. A connection between events on different hosts is established by connection events.

Definition:

> *A pattern* P, *relating events that occur at n distinct hosts, consists of n sets of events, one for each node. A set of events* S_A *at host* A *is linked to a set of events* S_B *at host* B, *iff* S_A *contains a send event and* S_B *contains the corresponding receive event. When node* A *opens a channel to* B *for data transmission (e.g. open a TCP connection, send a UDP packet, send an Ethernet frame), a pair of corresponding events (send at* A, *receive at* B*) is created.*

Definition:

> *Pattern* P *is valid, iff the following properties hold.*
>
> 1 *Each set of events is at least linked to one other set.*
>
> 2 *Every set except one (called the root set) contains exactly one send event. The root set contains no send event.*
>
> 3 *The connection graph contains no cycles. The connection graph is built by considering each event set as a vertex and each link between two sets as an edge between the corresponding vertices.*

These definitions actually only allow tree-like pattern structures (i.e. the connection graph is a tree), where the node with the root set is the root of the tree. Although this restriction seems limiting at a first glance, most desirable situations can still be described. Usually, activity at a target host only depends on events that have occurred earlier at several other hosts. This situation can easily be described by our tree patterns where connection links from those several hosts end at the root node. The opposite case, where events on two different nodes both depend on the occurrence of a single event at a third node is more difficult. In this case, the connection links do not end at the root node, but have their origin there. Such a situation cannot be directly expressed in our pattern language (as the root node set would contain two send events). Nevertheless, an application might split the original, illegal pattern into subpatterns (each representing a legal tree like structure) and relate the results itself. This allows to define arbitrary complex patterns at the expense of performance and network traffic. The major advantage of the proposed limitation is the possible implementation of an efficient search algorithm (for details, see Section 4) which transfers as little data as possible over the network.

Our query language allows to combine pattern specifications with the possibility to extract statistical data. A pattern instance is a set of events that satisfy the constraints of a valid pattern. Obviously, it might be possible that a single pattern is satisfied by more than one event set. Two event sets are said to be distinct, if they contain at least one distinct event element. An event element can be uniquely identified by its timestamp and the host, where it occurred. Statistical data can be computed for the set of all distinct instances of a given pattern. One can obtain the number of elements in that set (i.e. valid instances) or the maximum or minimum values for the number of instances at each host. Additionally, one can query attribute values of a certain single event of the pattern. The sum, maximum or minimum for an attribute may be calculated.

3.1. Event Query Language

This section describes the syntax and semantics of our Event Query Language (EQL) in more detail. We omit the complete language grammar, instead we gradually introduce the language by giving explanations on several examples.

A query is written as follows (similar to SQL).

```
SELECT results FROM nodes WHERE conditions
```

The *results* section is used to define the type of information the user is interested in. The operator `COUNT` can be used for patterns and returns a list of all nodes with the number (i.e. count) of found pattern instances at each one. The operators `SUM`, `MAX` and `MIN` may be used for complete patterns or for an attribute of a single event. When used for patterns, these operators return the sum, the maximum or the minimum number of detected pattern instances per node, respectively. When used for an event attribute, the sum or the extreme value (maximum/minimum) for a certain attribute value over all instances is returned.

The *nodes* section is used to assign an identifier to each node that is later used in the pattern definition. Additionally, one can impose restrictions on each node to have the agents only consider a limited set of actual hosts.

```
SELECT COUNT FROM host_1 range (10.1.17.0, 10.1.17.255)
# return the number of pattern instances for each host which is on the
# 10.1.17.* subnet (i.e. has an IP between 10.1.17.0 and 10.1.17.255
```

The *conditions* section specifies the pattern. It consists of a list of event sets, one for each node that appears in the node section. The event set is a list of identifiers, each describing an event. In order to be able to specify statistics operations on event attributes, one can assign identifiers (i.e. a label) to each of them. Two predefined labels called `send` and `receive` are used to identify the send and receive events, respectively, for linking event sets (see Section 3).

Each event can optionally be defined more precisely by constraints on the event's attribute values. These attribute values can be related to constant values or to variables by standard operators (`=`, `!=`, `<`, `>`, `>=` and `<=` with their usual semantic meaning) or by a `range` or an `in` operator as defined below.

x range $(x_0, x_1) \leftrightarrow x_0 \leq x \leq x_1$
x in $(x_0, x_1, ..., x_n) \leftrightarrow \exists i\ (0 \leq i \leq n)$ and $x = x_i$

A variable is defined the first time it is used. One must assign a value (bind an attribute value) to each defined variable exactly once while it may be used arbitrarily often as a right value in constraint definitions. The scope of variables is global and its type is inherited from the defining attribute.

With these explanations, we may introduce the syntax (in BNF) of the conditions section (all identifiers represent strings).

```
conditions   : {event set}+
event set    : node-identifier '{' {event}+ '}'
event        : [connection] event-identifier '[' {constraint ';' }* ']'
constraint   : assignment | [label] relation
assignment   : '$'variable-identifier '=' ( attribute-id | constant )
relation     : attribute-id operator ['('] {value ',' }* value [')']
value        : constant | '$'variable-identifier
operator     : '<' | '>' | '<=' | '>=' | '=' | '!=' | in | range
connection   : 'send('target-id'):' | 'receive('source-id'):'
label        : label-identifier':'
```

The following example shows a classical telnet chain pattern that describes a connection from Node1 to port 23 at Node2 and from there to port 23 at Node3. Node3 describes the root node set (i.e. has no outgoing send event).

```
Telnet Chain:

Node1 { send(node2): tcp_connect [] }
Node2 { receive(node1): tcp_accept [ port = 23; ]
        send(node3): tcp_connect [] }
Node3 { receive(node2): tcp_accept [ port = 23; ] }
```

4. Pattern Detection

Usually, patterns are discovered by gathering event data from distributed nodes at a central host where pattern matching algorithms distill the desired information. Our approach differs from the usual setup by detecting patterns in a fully distributed way using mobile agents. Mobile agents roam the network to search for suspicious events to start a more detailed investigation. When an agent spots the mark of a possible intrusion, it decides which data to carry with it on its next hop and which place to visit next.

The advantages of such a pattern detection scheme base on the fact, that no central entity is needed for data correlation. This increases the fault tolerance and robustness of the system, which is especially important for security relevant systems. When the host where a centralized IDS performs its correlation is taken out of action (e.g. by a DoS attack) the detection mechanism is actually blinded. When an attack renders some hosts in the network unavailable, agents can still search the remaining ones for signs of intrusions. Even when an attacker takes over a few hosts and manages to modify the agent platform in a way that it delivers wrong data (simply bringing it down is suspicious by itself), only intrusions where parts of the pattern occur at the compromised hosts are not detectable any more. The remaining system can still detect security violations.

Our approach also improves the scalability of the system because new hosts on the network won't automatically cause additional traffic to a single existing server machine. While traditional approaches like hierarchical installations and redundant servers allow to process more traffic than a single machine, a distributed approach is still desirable. We think that we can exploit the locality of network accesses. Most connections in large companies are between machines of the same department (like references to internal web servers or file shares) while connections between departments are rare. This allows agents to look for patterns in small areas and then move on. In a system with a central root node, all traffic would need to be forwarded (even with prefiltering and reduction over several hierarchies) to it.

The detection is done in the following way. An agent is started by the user interface with a given pattern (representing a security violation) that it has to look for. It starts its task by contacting the directory service to obtain a list of all hosts with an installed agent platform that match the constraints given in the pattern's `FROM` clause. These nodes are then visited in arbitrary order.

When an agent arrives at a host, it looks for events that fulfill the constraints given for the root node of the pattern it is currently investigating. In the case of the telnet chain introduced above, the agent would have to look for accepted TCP connections at port 23 (see Step 1 of Figure 2). The result of this process are a number of events (representing different instances of the pattern) which satisfy the root node constraint.

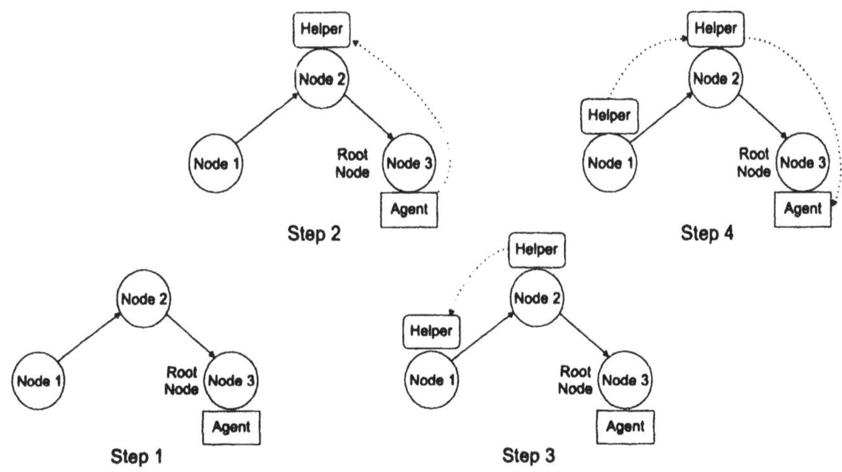

Figure 2. Pattern Detection

When no events are found, the agent immediately continues its journey. Otherwise, all receive events are identified. In the telnet chain case only receive events exist but more complex patterns are possible. For the distributed detection algorithm, a receive event is important because it establishes a relationship between nodes which is used to select a promising next place where an agent should look further. For each receive event, a helper agent is spawned which follows the link to the host with the corresponding send event (see Step 2 of Figure 2). By using the send event (determined by the receive event of the node the agent was coming from) it looks for events which fulfill the current part of the pattern. In the case of our telnet chain example, the agent already knows that the send event to Node 3 exists (as it is coming from there) and now searches for another accepted TCP connection at port 23 (from Node 1).

When the local event set contains receive events itself, the process recursively repeats by having the agent spawning helper agents and waiting for their return (see Step 3 of Figure 2). When the helper agents return, they report their findings (i.e. pattern instances) back to the agent waiting at the originating node (helper agents only move over a single hop). When pattern instances are returned, the waiting agent processes them (e.g. match variables or perform statistical evaluation) and eventually continues. When all helper agents have returned a pattern might be detected by the agent waiting at the root node as all information is available (see Step 4 of Figure 2).

Usually, only a small amount of data has to be transferred as it is not necessary to transport all pattern instances themselves but merely time stamps or single attribute values which have been assigned to variables.

Variables are treated in the following way. When a variable has already been bound to a value, it is straightforward to use this value directly for the attribute's constraint. This is the case when a value is assigned to a variable at a node which is closer to the root of the connection graph than the node, where it is used. On the other hand, when a constraint depends on a variable which has not been resolved yet, one has to temporary ignore that constraint. First, all actual bindings of the variable have to be determined and are then matched against the instances where the variable has not been assigned yet. This allows to filter out pattern instances which do not satisfy the previously ignored constraint. Notice that this may cause agents to transfer unnecessary pattern instances but the runtime complexity is still linear with regard to the number of send/receive events.

Instead of having a single agent visit all nodes sequentially, the task could be parallelized easily by partitioning the set of interesting hosts. Each partition is visited by a dedicated agent which all have to agree on a destination node, where they meet and merge their results.

We have installed a first prototype version at our department's network as universities are traditionally favorite targets of hackers. We are currently able to detect about a hundred local events (by looking for well known attack signatures and network connections) and a dozen distributed patterns. The results are promising as a couple of incidents have already been detected. The network overhead of the traveling agents is negligible and the processing overhead at each node is reasonably low.

5. Security

Mobile code introduces a number of security issues that our design has to deal with. Especially when building systems for security sensitive applications (like our intrusion detection system), it is important not to introduce new vulnerabilities by the security monitoring tool itself. The security threats to mobile agents are classified by four categories, namely agent-to-agent, agent-to-platform, platform-to-agent and other-to-agent.

- *Agent-to-agent threats* describe the set of attacks, where one agent exploits the vulnerabilities of another agent. In Sparta, agents only locally communicate with helper agents they have previously spawned. As arbitrary interagent communication is prohibited, possible vulnerabilities cannot be exploited and agent-to-agent attacks can be prevented.
- *Agent-to-platform threats* describe attacks, where an agent performs malicious activities against its environment (i.e. platform). To prevent these kind of attacks, the runtime permissions of agents are rigorously restricted. They are not allowed to access resources directly. Instead, agents gain information by querying the data storage component.
- *Platform-to-agent threats* describe situations, where the platform compromises the agent's security. This sort of threats is extremely difficult to

defend against when agents need unrestricted movement around the network (Jansen and Karygiannis, 1999) and Sparta has no special mechanisms to defend against such attacks. In contrast to a central server system, we still have the advantage that even if a single node is compromised, all patterns which do not touch this host are still detectable.

- *Other-to-agent attacks* involve threats against agents performed by external entities while they are in transit over the network (e.g. eavesdropping or tampering). Sparta uses an asymmetric (public/private key pair) cryptosystem to secure agents when they are transferred over the network. The agent code is signed and can be authenticated before it is executed. In order to manage the asymmetric cryptosystem, a Public Key Infrastructure (PKI) is provided.

6. Related Work

The idea of correlating events which occur at different places in a network and to formalize patterns to describe such correlations is not new. The Complex Event Processor which is developed at Stanford University is capable of correlating causally and temporally related events. It bases on the theory of partial ordered multisets (Pratt, 1986) and is used for intrusion detection (Perrochon et al., 2000) and network management (Perrochon et al., 1999). Patterns are described using the Rapide Pattern Language (RAPIDE, 1997). The difference between our approach and their work is the fact, that we use mobile agents to perform the pattern detection in a distributed fashion without any central server. In contrast to that, they collect data from different client sites and process it at a server. Rapide has clearly influenced our work as their system as well as ours try to correlate generic events and target a broad spectrum of applications.

State-of-the-art ID systems like EMERALD (Porras and Neumann, 1997), NStat (Kemmerer, 1997) or AAFID (Balasubramaniyan et al., 1998) can gather and relate data from different sources. In contrast to our distributed design, they have a hierarchical architecture where sensors located at different hosts collect data and send it to a central entity where events are related. Unlike Rapide and our design, they completely focus on intrusion detection events and are not applicable to different domains. The same is true for network management software (e.g. HP OpenView (Sheers, 1996)).

Commercial intrusion detection systems like Network Flight Recorder (NFR, 2001) or Real Secure (RealSecure, 2001) perform their analysis on packet level by monitoring network traffic. This allows only simple correlation, but their output can be used as our basic events.

IDA (Asaka et al., 1999) uses mobile agents to trace a possible attacker back to its origin, while Micael (de Queiroz et al., 1999) pursues a more ambitious aim where each system component is realized as a mobile agent. Unfortunately, only a high-level system design has been presented. The possible advantages of mobile agents in intrusion detection systems are summarized in (Jansen et al., 1999) and (Krügel and Toth, 2001).

7. Conclusion

Relating distributed events and deducing knowledge from different hosts is especially important in the field of network management and intrusion detection.

We present a solution, where mobile agents perform the task of correlating data in a fully decentralized manner. In order to prevent a tremendous increase in network traffic, the expressiveness of our pattern description language had to be slightly restricted. This allows an efficient detection algorithm and a fault tolerant and scalable system design.

References

Asaka, M., Taguchi, A., and Goto, S. (1999). The implementation of ida: An intrusion detection agent system. In *Proceedings of the 11th FIRST Conference.*

Balasubramaniyan, J. S., Garcia-Fernandez, J. O., Isacoff, D., Spafford, E., and Zamboni, D. (1998). An architecture for intrusion detection using autonomous agents. In *14th IEEE Computer Security Applications Conference.*

de Queiroz, J. D., da Costa Carmo, L. F. R., and Pirmez, L. (1999). Micael: An autonomous mobile agent system to protect new generation networked applicatiohs. In *2nd Annual Workshop on Recent Advances in Intrusion Detection.*

Jansen, W. and Karygiannis, T. (1999). Mobile agents and security. Special Pub. 800-19, NIST.

Jansen, W., Mell, P., Karygiannis, and Marks, D. (1999). Applying mobile agents to intrusion detection and response. Interim Report (IR) 6416, NIST.

Kemmerer, R. A. (1997). A model-based real-time network intrusion detection system. Technical report, Computer Science Dep., University of California Santa Barbara, November.

Krügel, C. and Toth, T. (2001). Applying mobile agent technology to intrusion detection. In *ICSE Workshop on Software Engineering and Mobility.*

NFR (2001). Network Flight Recorder. http://www.nfr.net/.

Perrochon, L., Jang, E., and Luckham, D. C. (2000). Enlisting event patterns for cyber battlefield awareness. In *ARPA Information Survivability Conference and Exposition (DISCEX'00).*

Perrochon, L., Kasriel, S., and Luckham, D. C. (1999). Managing event processing networks. Technical Report CSL-TR-99-877, Stanford Computer Systems Laboratory.

Porras, P. A. and Neumann, P. G. (1997). Emerald: Event monitoring enabling responses to anomalous live disturbances. In *Proceedings of the 20th NIS Security Conference.*

Pratt, V. (1986). Modelling concurrency with partial orders. *Int. Journal of Parallel Programming*, 15(1):33–71.

RAPIDE (1997). *Rapide 1.0 Pattern Language Reference.* Stanford University.

RealSecure (2001). RealSecure. http://www.iss.net/customer_care/resource_ center/product_lit/.

Sheers, K. R. (1996). HP OpenView Event correlation. *Hewlett-Packard Journal.*

Staniford-Chen, S., Cheung, S., Crawford, R., Dilger, M., Frank, J., Hoagland, J., Levitt, K., Wee, C., Yip, R., and Zerkle, D. (1996). Grids - a graph based intrusion detection system for large networks. In *Proceedings of the 20th National Information Systems Security Conference*, volume 1, pages 361–370.

PART TWO

Invited Papers

SHELL'S TRUST DOMAIN INFRASTRUCTURE SECURITY CERTIFICATION

Linking security management to business objectives

Pieter van Dijken
Shell Services International
Piet.P.vanDijken@IS.shell.com

Abstract Shell companies worldwide completed in 2000 a security programme, covering certification of their IT infrastructure against a subset of ISO 17799 and related ISO standards on certification and audit. Objective was to provide the Shell Group with a secure environment to do ("e") business in, i.e. sharing of knowledge, enabling support for global applications and supporting virtual teamworking. The scheme is now up and running in more than 160 countries and 240 Shell companies. In this presentation I will describe background considerations on the Scheme as an example of business linked information security management. I will go as well into practical issues regarding roll-out and implementation of a global scheme like this. I will conclude with outlook for the Scheme, planned activities and issues.

Pieter van Dijken (51, Dutch) manages the global information security consultancy team in Shell Services International. This team operates from locations around the world in support of the information security requirements of the Royal Dutch/Shell Group of companies. His team facilitated very recently a strategic, world wide security certification programme, called Trust Domain. Objective of this programme is to establish a common set of information security standards and controls for IT infrastructure throughout the Shell Group, based on a subset of the BS 7799 standard. Benefits of having such a common set are numerous, e.g. facilitate information sharing across the Shell Group without making Shell companies vulnerable to unauthorised access; having lower cost and more capabilities through the use of standard security protocols and tools and finally by avoiding unnecessary security controls between Shell companies.

Pieter joined Shell in 1988 and has been involved in numerous international policy and standardisation efforts with regard to trust and confidence in IT since. He was directly responsible for the translation of BS 7799 in Dutch, took part or chaired a host of related initiatives (e.g. the Dutch BS 7799 certification Scheme and many others). Pieter has degrees in business law and police administration and published on criminal justice implications of IT. He lives in the Netherlands with his wive and three children.

AUTHOR INDEX

GPSR Compliance
The European Union's (EU) General Product Safety Regulation (GPSR) is a set of rules that requires consumer products to be safe and our obligations to ensure this.

If you have any concerns about our products, you can contact us on

ProductSafety@springernature.com

In case Publisher is established outside the EU, the EU authorized representative is:

Springer Nature Customer Service Center GmbH
Europaplatz 3
69115 Heidelberg, Germany

www.ingramcontent.com/pod-product-compliance
Ingram Content Group UK Ltd.
Pitfield, Milton Keynes, MK11 3LW, UK
UKHW012211240726
13966UKWH00002B/696

* 9 7 8 1 4 7 5 7 7 6 6 4 5 *